P9-AON-580

NARRATIVE
EXPERIMENTS

No Longer Property of
Phillips Memorial Library

NARRATIVE EXPERIMENTS

The Discursive Authority
of Science and Technology

Gayle L. Ormiston
and Raphael Sassower

University of Minnesota Press • Minneapolis

PHILLIPS MEMORIAL
LIBRARY
PROVIDENCE COLLEGE

Copyright © 1989 by the Regents of the University of Minnesota

All rights reserved. No part of this publication may be
reproduced, stored in a retrieval system, or transmitted, in any
form or by any means, electronic, mechanical, photocopying,
recording, or otherwise, without the prior written permission
of the publisher.

Published by the University of Minnesota Press
2037 University Avenue Southeast, Minneapolis, MN 55414.
Published simultaneously in Canada
by Fitzhenry & Whiteside Limited, Markham.
Printed in the United States of America.

Library of Congress Cataloging-in-Publication Data

Ormiston, Gayle L., 1951–
    Narrative experiments : the discursive authority of science
    and technology / Gayle L. Ormiston and Raphael Sassower.
        p.   cm.
    Includes bibliographical references.
    ISBN 0-8166-1820-8. — ISBN 0-8166-1821-6 (pbk.)
    1. Science—Philosophy.   2. Technology—Philosophy.
    I. Sassower, Raphael.   II. Title.
    Q175.068   1989
    501—dc20   89-39005
                                                                CIP

The University of Minnesota
is an equal-opportunity
educator and employer.

– for Anna and Michael
GLO

– for Helena Sassower
RS

CONTENTS

PREFACE

We explore the cultural/linguistic matrices in which *science* and *technology* take shape. From the outset, we do not assume that science and technology constitute domains of inquiry separate from the inquiries undertaken in other traditional disciplines or areas of research, such as philosophy, aesthetics and literary criticism, music, or hermeneutics (the histories and theories of interpretation). Instead, we assume that, like any other inquiries, science and technology are conditioned by particular etymological and historical requirements, as well as specific interpretive and pragmatic demands. The issues typically associated with scientific investigation and technological innovation can no longer be comprehended as limited to the domains of *science* and *technology*. Instead, they are understood within what we call the labyrinths of cultural and linguistic usage. Stated in a more direct manner, there are no issues of science and technology as such. Thus, this text is itself an *experiment*—an experiment in narration.

One purpose of this text is to undertake, explicitly, an exploration of narratives and texts that have been included within the domains of science and technology, and that is itself a textual experiment. At one point, we trace the overlay or interlacing of certain lines of thought that design and prefigure the interpretive analyses or critiques undertaken, and the verdicts these critiques render, in the name of any particular world view (Weltanschauung). In other words, our approach to comprehensive systems (like those presented by Bacon, Kant, or

Rorty)—that is, narratives that presume a correspondence between their representations and an "actual state of affairs"—is to engage them as nothing more nor less than *counterfeits*. Each system is cast as an attempt to remake or redesign the world *counter* to what it interprets as the dominant model or narrative. That is to say, each and every interpretation presents itself as a replacement that overcomes the deficiencies and incompleteness of other interpretations.

At another point, we stress the *performative* character of interpretation. Each interpretation, and as such each narrative or systematic account, is carried out according to rules that are first declared and, thus, legitimated in and through its performance. They are not, then, rules for which there is any support or grounding external to the context of a performance and the performance itself. Like any interpretation, text, or narrative, the interpretations we offer in this book recall and reiterate that interpretation is an intervention that results in discursive displacement and not replacement. No replacement of one interpretation by another, of one theory, model, or text with another is ever complete or final. Instead, the performance or the activity of interpretation is the re-collection and re-iteration of certain themes, problems, issues, and questions already articulated and advanced—on some stage or platform—within the labyrinths of narrative accounts.

And at yet another point, we draw attention to the recurrence of certain themes central to inquiries concerned with the nature and scope of science and technology. The prominence of these themes is interpreted in terms of how their specific textual uses produce certain conceptual and practical bifurcations within the discourses on science and technology. The opposition of theory and practice, science and technology, certainty and ambiguity, foundation and abyss, or language and nature, for example, determines the limits of the discourses we examine. Yet, the recurrence of these and other themes shows how discursive boundaries—whether one thinks of these boundaries as disciplinary or textual—undergo constant transformation with each interpretation.

But no re-collection is the repetition of the same. Re-collection involves thematic displacement. The themes and ideas and problems reiterated here are cast in a context different from the context of their initial presentation. Here use within a specific context differentiates one orientation from another. Because we experiment with the concepts of authority, practice, and textual (or narrative) labyrinths, this book focuses on pedagogical issues. Because we experiment with the notions of context, use, and discursive conditions, this book focuses on pragmatic interpretive devices. Because we experiment with etymologies, interpretive techniques, and linguistic (re-)presentations (fictions), this book focuses on textual criticism. Because we experiment with the ideas of enlightenments, legacies, and cultural domains, this book focuses on the history of ideas and systems of thought. And, because this book focuses on the practices of discursive displacement and the dissemination of authority, we experiment with the notions of *science* and *technology*.

Because we experiment, because the narratives of science and technology are experiments themselves, and because the results of no experiment can be predetermined nor interpreted in a definitive fashion, this book does not present another "grand" or "Meta" narrative. Like all other experiments, narratives, or interpretations, it presents interpretations of interpretations, narratives about narratives, and critiques of critiques. The only privilege accorded our account is the privilege accorded any other *me*tanarrative that situates itself in between and mediates other narratives.

There are many stories we could tell about the many people who have been involved in the various stages of this experiment. Instead of recalling the details of their involvement, we would like to acknowledge our debt to them by name. For institutional support: Hunter R. Rawlings III, Dwayne C. Nuzum, Joan Klingel Ray, James A. Null, and the University of Colorado Committee on Research and Creative Works. For scholarly comments and suggestions: Paul Durbin, Larry Hickman, and Gerald Kreyche. For their unwavering intellectual support and personal encouragement, we thank Mary Lynn Ormiston and Galit P. Sassower. And, finally, we acknowledge with gratitude the technological and scientific innovations that made our collaboration possible.

NARRATIVE EXPERIMENTS

CHAPTER 1
THE INTERPLAY OF SCIENCE AND TECHNOLOGY
An Introduction

The Linguistic Context

The exploration of what is called "science, technology, and society" assumes an interlacing of the multiple cultural and linguistic dimensions of contemporary life.[1] That is to say, the examination of the interplay of science, technology, and society presumes that *all inquiry* presupposes and unfolds within a cultural context that always determines and is already determined by social relationships. The techniques and strategies of inquiry, the *methods* used to identify so-called "significant" problems or questions facing a particular group or generation of individuals, are conditioned and rendered possible by the current historical, material, and intellectual culture.

Even though the field of this inquiry is diverse, the study of the interplay between science, technology, and society does not constitute a field or a discipline. Scattered across traditional disciplinary boundaries, it transforms the concerns and questions, the methods and techniques of inquiry that identify and secure disciplinary inquiry by situating them within a different context. No longer are there "philosophical" issues, "technical" questions, "moral" problems, or "scientific" hypotheses as such. These issues are comprehended according to their placement within a chain of other overly determined and supplementary issues and questions.

Within different historical and cultural contexts, words are used for different purposes of designation and, as such, their meanings differ. Use creates. In the case of signs, symbols, or words, use creates meaning.[2] It orders, in each instance, a different concentration of significance or value, according to the political, psychological, theological, philosophical, and scientific topography in which a word is deployed. One way of undertaking an analysis of the broad cultural matrix that informs the development and study of *science* and *technology*, then, is to reflect on the history of each term's usage. Etymological deliberations offer a provisional means by which the current use of a term can be comprehended according to its placement within a series of other uses and significations already attached to it but, nevertheless, sometimes overlooked or forgotten. If one prefers, etymological reflection elicits a "recollection" (*anámnēsis*), analogous to that discussed by Plato in the *Meno* (81 c-d), of what is already presupposed and buried by a word's contemporary use. Etymological and lexical deliberations trace histories and, as such, indicate ways of comprehending subtle differences in use and deployment.

According to its rendering in *The Oxford English Dictionary*, *science* stems from the Latin *scientia* meaning "knowledge," a participle form of *scire* meaning "to know." According to its Latin roots, then, *science* designates the "state of 'knowing'" acquired by the study and mastery of a specific discipline, a particular branch of knowledge. Here *science*—the word and the *activity*, it should be noted—is used in a more restricted fashion. It is identified with "a connected body of demonstrated truths or with observed facts systematically classified and more or less colligated by being brought under general laws . . ." In its more recent determinations *science* develops "trustworthy methods for the discovery of new truth within its own domain." Thus, *science* is associated with a certain kind of knowledge or learning, one that is distinct from art, as a craft, trade, or skill—*techné*.

But the juxtaposition of *science* and *techné* shows that the two activities were not always so dramatically opposed to one another. According to the Greek roots, *epistémé* (true knowledge) is concerned with "theoretic truth" whereas *techné* (art) is concerned with "methods for effecting certain truths." Where *epistémé* would be extended to denote "practical work which depends on the practical application of principles," *techné* would designate art that required "knowledge of traditional rules and skills acquired by habit." In some instances, in the context of Greek reflective thought, science and art—*epistémé* and *techné*—are used interchangeably. Theoretical knowledge requires practical application of principles; art presupposes knowledge of rules. Moreover, there are moments within this etymological context where *science* refers to a "craft, trade, or occupation requiring trained skill."

In these latter determinations, *epistémé* and *techné*, or *science* and *technology,* are already interdependent. Knowledge of ideals, principles, and rules is knowledge derived from practice, from the need to satisfy a practical demand. In an analogous fashion, the engagement of ideas within a specific context is functionally dependent on an awareness of concepts or rules derived from the habit of performance. One might imagine this interplay in terms of what Gilbert Ryle calls the relation between "knowing-that" and "knowing-how."[3] Once again, an appeal to etymology can assist in understanding this interplay.

As already noted above, *scientia* is a participle form of *scire*. *Scire* has its roots in *skei*, which means "to cut" or "to split." Knowledge, then, is understood as the *ability*, the *skill* "to separate one thing from another," "to discern." In the Greek, such separation is related to *skhizein* meaning "to split" into many parts, which is the root for *schizo-* and *schism*. Thus, the ability to discern differences, or what Plato calls in *The Republic* a certain kind of "mindfulness" (621 c), is related to another Latin root *skel*, which also means "to cut" but which is more directly related to a concept developed later in the Old Norse where *reason, knowledge,* and *incisiveness* are comprehended by *skil*, a precursor to our contemporary term *skill*. *Scire, skei,* and *skil* are derivations of *sek*, a basic form of the Latin verb *secare*, meaning "to cut." But here the separation, the segregation of knowledge is not so acute. *Sek* is also "dissect," "exsect," "intersect," "notch," "resect," and "transect."

The point of this momentary etymological reflection is to show that the pedestrian separation of *science* and *art*, or *science* and *technology* is a function of the purposes—the theoretical and practical ends—to which these words and activities are directed within certain historical contexts. But what is more pertinent to the purposes of this introductory discussion is how the subtle differences in meaning, the differences initiated through carefully nuanced articulation, already presume and incorporate *the artifice of usage,* that is to say technology. Thus, *discernment,* or what the Greeks understood as "scientific knowledge," presupposes some systematic appreciation of means and skill, means by which something is created, such as the categorization of knowledge. The ability to discern is a condition for the possible separation of science and technology. Now we can turn to a brief reflection on the etymology of *technology, techné-logia*.

Appealing to *The Oxford English Dictionary* once again, we find that a gloss of its renderings reveals that *technology* stems from the combination of two Greek terms, *techné* and *logos—techné-logia*. *Techné*, as indicated above, designates art, craft, skill, a way or manner, a means whereby something is contrived, created, or devised by art. *Logos* is rendered as discourse, account, definition, and proportion (as an organizational principle). Combined, *techné-logia* is a discourse or a treatise on an art or arts; the systematic treatment of the arts, sometimes the practical arts collectively. Here the systematic account is transformed into theoretical discourse. More recent entries cite slight alterations and transformations of the term's usage: technology is identified with "weaving," "cutting canoes," and making "rude weapons." It is not until the later part of the nineteenth century that technology is linked with "applied science."

The reference to the etymologies of *science, knowledge,* and *technology* is not to be construed as an effort to imitate, to copy, or to reclaim a lost or forgotten meaning. Instead, tracing the antecedent derivations of these terms helps us understand the difficulty in providing a precise definition of *science* and *technology*. All the predicates, all the defining concepts, all the lexical significations, which seem at once to fall under a particular rubric, are conditioned by and products of the interplay between science and technology. As the etymologies suggest, the value of words, concepts, and activities—in this case, *science* and *technology*—is bound to a specific context, a cultural/linguistic matrix that is unavoidably composed of ethical, political, artistic, literary, psychological, sociological, and scientific interests.

Science versus Technology or Science/Technology

To stress the etymologies of *science* and *technology* involves more than what might be termed a mere lexical understanding of their relationship. The histories of these two terms are interpreted within certain cultural contexts. As such, these histories inform the traditional views concerning the interplay of science and technology. Our interest and concern with this question arises in response to how the interplay has been construed or depicted in prevailing philosophical and literary accounts of Western thought.

There is, to be sure, the problem of distinguishing between the kinds of activities that are identified as scientific and technological. In some cases, common understandings of science and technology assume that *technology* refers to "scientific technology," or "high-level, big technology," and as such is related to science in the sense of attempting to apply scientific theories.[4] The implementation, then, becomes the basis for making claims regarding the "validity" of these theories. But, given what has been suggested in the return to an etymological context, is this not a rather restricted use and parochial understanding of the terms?

As we have seen, *technology* designates daily activities, trades, and crafts supposedly not associated with any scientific theories. Moreover, it signifies the management of the political arena, as reflected in Plato's treatment of Socrates' concerns with the establishment of the polis,[5] and domestic (economic) affairs of households, as reflected in Xenophon's works.[6] These two realms traditionally fall under the classification of specific sciences or theories.

To the extent that the use of the terms *science* and *technology* already constitutes an interpretation, and that interpretation is informed culturally and linguistically (that is, its significance remains openended), then it seems their interplay cannot be characterized in any petrified manner. One way of comprehending the interplay is to review, very briefly, how it has been articulated in the history of Western thought by two prominent schools—classical idealism and classical materialism.

Up to the end of the nineteenth century, technology had been cast as the "handmaiden" of science by classical idealism. Such an image presumes the superiority of science over technology according to a hierarchy of human endeavors, where the domain of science is theoretical and the domain of technology is practical. According to this view, technological experiments and innovations are nothing more than the application of certain theoretical formulations of science.

Opposed to the idealist's perspective, regarding the subordinate role technology plays vis-à-vis science, there is a competing view aligned with classical materialism. According to the materialist point of view, technology, understood in terms of techniques and practices, precedes science chronologically and logically. Chronological priority deals with daily practices and routines handed down by tradition and refined by groups who need certain techniques for their daily survival.

There are numerous examples in the history of Western thought that seem to support the priority of technology over science. It is common to advance the claim that the theoretical developments in geometry were due to land measurements in the Nile (because continuous floods erased the borders between farms) and to the stellar observations of marine navigators. Similar claims have been made concerning the origins of modern probability theory. The practical concerns of insurance companies in England (of having to insure trading ships that were going as far as India) as well as the concerns of gamblers in

France, motivated some mathematically competent thinkers to develop a theoretical formula that projected the probabilities of survival or of success. When practices and techniques eventually lead to some theory, scientific or not, its longevity depends on the continued "empirical" support it receives from these practices, these techniques.

To supplement the belief in chronological priority, and to ward off the criticism that this priority is accidental, a logical priority is assigned to technology. One may associate this second sort of priority with classical Baconian inductivism. Through the accumulation of numerous accounts and testimonies regarding a set of practices and techniques, we reach a level of generalization that permits us to construct a theory, now understood as a scientific theory proper. When theories are said to correspond to actual experiences, they are more likely to support claims of meaning and validity.

Neither classical idealism nor classical materialism appreciates the complexity and richness of the interplay between science and technology indicated by the return to their etymologies. Though each approach provides convenient images and metaphors by which we can approach the history of the relationship itself, it seems that to grasp the intricacies of this relationship both perspectives must be engaged. Can it be said, then, that these two views complement each other? That is, can both views be entertained simultaneously?

Instead of examining the interplay of science and technology from these classical perspectives, we suggest that the relationship needs to be rethought, or that a new orientation must be provided. This end can be achieved if attention is redirected towards the relationships that can be drawn between science, technology, and the humanities, broadly understood. Such a rethinking must recognize the difficulty of demarcating science and technology in an unambiguous manner except for the sake of linguistic or practical convenience.

We do not recommend a return to the Greek conception of the overlap between science and technology. We would like to leave open the question as to whether or not in the late twentieth century the convergence *or* divergence of science and technology is to be emphasized. Arguments can be articulated and supported to show that science acquires its value and foundation through the employment of technological tools. By the same token, for policy purposes it may be prudent to distinguish between the open-ended pursuits of theoretical physics, for example, while restricting the implementation of nuclear physics in the form of power plants. These examples illustrate the need to appreciate the interplay of science and technology in terms of the cultural context in which it takes place.

The cultural context is not merely a background for the examination of the historical development of science and technology, taken separately or in terms of the intersection of certain areas of human activity. Instead, the activities themselves are responsible for the culture that engenders and nurtures them. Here the difficulties of simple demarcations are more pronounced. One cannot provide an account that captures the "real" causal or logical connections that might exist between science and technology and the humanities, or the reverse. In general, our views of science and technology are embedded in the words we use to talk about them. In certain cases, this discourse becomes theoretically rigid, resulting in the jargon(s) of science and technology. Thus, this kind of theoretical rigor creates a breach in the relationships between the discourses of science, technology, and the humanities.

These gaps exacerbate the difficulty of presenting the formal content of what is considered today as the study of science, technology, and society. The interplay between science and technology takes on a different meaning if it is understood in pedagogical terms.

Linking Theory and Practice

Concerns over technical education are neither new nor limited to our culture or historic moment. The eleventh edition of the *Britannica* (1911) contains an interesting discussion of technical education by Sir Philip Magnus, who was among other official duties, a member of the Royal Commission on Technical Instruction from 1881 to 1884. Magnus notes a difference between modern technological teaching and the mechanical practices of manufacturing and commerce. Teaching science and technology in schools is different from the apprenticeship method employed for the acquisition of specific skills. But what constitutes the difference? According to Magnus, students "should be so taught as to become instrumental in the formation of mental habits and the development of character, the mere knowledge or skill acquired being of secondary importance."[7]

Magnus's remarks apply in a limited fashion to the interplay of science and technology in cultural terms. We cannot ignore the difficulty of determining what is to be included in such studies, and devote our attention exclusively to the means by which we present the material to others—especially students. Besides, do we need to elaborate any etymological discussion of science and technology as subject matter, if our concern, according to Magnus, is indeed the development of mental habits and character in our students, or what amounts to the development of technical skills for properly conducting our thoughts and actions?

An American contemporary of Magnus, Charles Sanders Peirce was also interested in the pedagogical difficulties and methodological issues associated with the clear presentation of ideas. After a brief review of Descartes's recommendation of the a priori certainty of specific knowledge claims and Leibniz's insistence on the close scrutiny of the definitions we use when we speak of our knowledge claims, Peirce felt it necessary to suggest an alternative. Dissatisfied with what he took to be Descartes's conviction for a priori truths concerning self and God (a conviction fundamentally pleasing to reason), and suspicious that Leibniz's recommendation would not lead beyond the clarification of linguistic use, Peirce put forward another method for determining the clarity of ideas. He formulated the *pragmatic maxim*: "consider what effects, which might conceivably have practical bearings, we conceive the subject of our conception to have. Then, our conception of these effects is the whole of our conception of the object."[8]

Peirce's advice is more useful for present purposes than is Magnus's. Peirce's maxim challenges the very conceptions we may have of our material, convictions that might be pleasing to reason or mere theoretical considerations, speculations that have no bearing on practice. Instead of thinking about science and technology in exclusively "theoretical" terms, we must incorporate our theoretical concerns (like the Greeks perhaps) with other issues of practice. The clarity of perception, and, therefore, the clarity of presentation, depends on the appreciation that ideas make sense only when they are placed into action, that is to say, comprehended according to the conditions of a specific context. The linkage between the theoretical and practical constitutes the discursive framework in which any pedagogical discussion concerned with the character of the culture it formulates must take place.

Our brief reflection on Peirce's pragmatic maxim provides the opportunity to state explicitly a presupposition that informs Peirce's methodological considerations and underlies our efforts in this book. Theory—conceptualization—is born in practice, refined through inquiry (the *practice* of putting ideas into play as hypotheses, so to speak), and returns to practice in our attempts to realize the differences our conceptions make in the interrelation with other ideas.

Moreover, this reflection draws attention to a specific problem facing any understanding of the role and function of science and technology in our culture. If the study of science and technology is to contribute to the development of "mental habits," to use Magnus's term, or if it is to bring about a certain kind of "critical awareness" or "enlightenment" regarding the ubiquity of technology, then it must make a *dif-ference* in action. It must generate some change. According to Peirce, "the conceivably practical" is the "root of every real distinction of thought, no matter how subtle it may be; and there is no distinction of meaning so fine as to consist in anything but a possible difference of practice."[9] What is the general—pedagogical, epistemological, metaphysical, political, discursive—significance of the discussion that focuses on the interplay of science and technology?

The Conceptual and Practical Context of Current Issues

To portray the interplay of science and technology within specific lin-guistic and cultural contexts is not merely another interdisciplinary ap-proach to this interplay. Of course, there are several ways to conceive of interdisciplinary studies. One way would be to emphasize the os-tensible overlap of different disciplines, where the methods of inquiry typical of specific disciplines are joined in the study of a specific object or problem. One can imagine this situation with political philosophy, political science, and sociology; aesthetic theory, the history of art, and literary criticism; or molecular biology, physiology, and organic chemistry.

A second way might be to focus on science and technology as the objects of inquiry from a fusion of several disciplinary perspectives. In this way, one might focus on only the political aspects of science and technology; the challenge to theological beliefs due to the increasing changes in our lives effected by advances in science and technology; or the ethical consequences pertaining to the use of certain scientific and technological innovations.

A third way might be to take science and technology as the founda-tion for examining the contemporary status of a specific discipline. For example, one can imagine examining certain styles of contemporary art on the basis of the technologies employed to generate certain aes-thetic effects and illusions. In a similar fashion, the advance of political science or physical geography could be examined in terms of their de-pendence on the actual and potential use of computer technology.

The most prominent attempts to overcome disciplinary boundaries, or to become interdisciplinary, in at least one of the senses described above, may be represented in the following manner.

Defining Technique

Jacques Ellul, in *The Technological Society*,[10] insists upon the critical reflection on the difficulties involved in arriving at a working definition of *technique*, technology. Ellul insists that we discard our traditional conceptions of technology, where it is understood simply in terms of craftsmanship, tools, machinery, and practical skills. Instead, technology has assumed a cultural role, creating its own mentality.

The Autonomy of Technology

Written in a genre analogous to Ellul's, Langdon Winner's *Autonomous Technology: Technics-Out-of-Control as a Theme in Political Thought*[11] provides another attempt to understand technology apart from its mechanistic apparatus. Winner challenges the claim that the threat of technology lies in its having become an autonomous agency, independent of human control. Rather, Winner claims that the question of the autonomy of technology must be understood as a political phenomenon, reasserting the problem of human autonomy.

Critiques of Utopian/Dystopian Perspectives

Like Winner, Bernard Gendron, in *Technology and the Human Condition*,[12] casts technology into the political domain. Within this spectrum, Gendron provides one of the earliest, sustained socialist critiques of the utopian and dystopian views of technology. His socialist analysis is in terms of the exploitation and alienation of labor, the power relations between those owning and those using technology, and the emancipatory potential of technology.

Controlling Technology

Designed specifically for use in a course titled "Control of Technology," offered at the Open University of London, Godfrey Boyle, David Elliot, and Robin Roy edited *The Politics of Technology*[13] as a presentation of a series of readings selected to cover a broad interdisciplinary approach, emphasizing the divergent views about how the social control of technology can be best achieved. This collection's significance lies in its attempt to bring the social and political context of technology to the classroom as an object of examination.

Philosophical Examinations

With the Carl Mitcham and Robert Mackey collection, *Philosophy and Technology: Readings in the Philosophical Problems of Technology*,[14] the field now known as "the philosophy of technology" is given an extensive survey as it has evolved into a subject of general (philosophical) interest. The survey it contains proceeds from a variety of philosophical orientations that address specific questions related to the cultural emergence of technology.

The Status of Technology in Everyday Life

Following in the phenomenological tradition of Husserl and Heidegger, Don Ihde, in *Existential Technics*,[15] presents an existential analysis of the interaction between human beings and the technologies (technics) that inform their self-understanding. Ihde argues that technology is "historically and ontologically prior to science." To articulate this unique perspective, Ihde must return to a description of technology that stresses its variegated appearance in our daily experiences.

Feminist and Gender Critiques

Joan Rothschild's collection, *Machina ex Dea: Feminist Perspectives on Technology*,[16] exemplifies current application of feminist critiques in the field of philosophy and technology. It explores the historical, social, and psychological reasons behind the absence of any appreciation of the role feminist perspectives or gender issues play in the overall attempt to understand the development of technology. Overlooking the important role women have played in the preservation, development, and innovations of specific technics (e.g., associated with household economics), the philosophy of technology has been theoretically disabled—it lacks the breadth and scope with which to challenge traditional views of technology.

Ethical Considerations

Hans Jonas's contribution, in *The Imperative of Responsibility: In Search of an Ethics for the Technological Age*,[17] lies in his recognition that modern technology has changed our natural surroundings as well as our social fabric and individual perception of ourselves. This change is so radical that it is no longer feasible to talk about the control of technology or the containment of specific technological hazards. Instead, Jonas demands that we engage in a reevaluation of technology in terms of our own eminent fear of it, so as to create an ethical framework for responsible coexistence with it.

Social and Political Concerns

The significance of Joseph Agassi's *Technology: Philosophical and Social Aspects*[18] lies in his attempt to articulate a democratic framework for the social and political control of technology. Focusing on a variety of topics, Agassi illustrates the philosophical presuppositions that influenced the social acceptance or rejection of technological innovations. The significance of Agassi's work lies in the democratic and "rationally-based" recommendations it offers regarding the role of technology in the conduct of contemporary life. In this regard, it proffers an alternative to Jonas's appeal to the overwhelming threats posed by modern technology.

Science/Technology as Language/Culture

Rather than thinking of the interplay as yet another *inter*disciplinary or *trans*disciplinary or *multi*disciplinary topic, it seems that a new designation as a "metadiscipline" is needed. The importance of the *meta* distinction rests on the realization that the reciprocity of science, technology, and the humanities obliterates all traditional disciplinary boundaries. Reciprocity takes place between these boundaries, where no one method, conceptual apparatus, or scholastic conviction succeeds on its own. Thus, there is no object per se, no "philosophy of technology," set apart from the interrelation of thought, language, practice, and culture. If such a reciprocity is to be appreciated, it must be seen as an erasure of the boundaries and the intellectual segregation of disciplinary inquiry. It is an erasure simultaneously groping to establish new limits.

The metadisciplinary perspective of this text offers a critically comprehensive overview of the cultural and humanistic context of science and technology. Such an account is not concerned to provide a hierarchical ordering of science, technology, and the humanities. Instead, it attempts to undermine any such ordering by demonstrating how science, technology, and the humanities develop in concert with one another; they are mutually constitutive of one another and their culture. Science, technology, and traditional humanistic studies, then, are modes of one another.

Four themes are developed in the following chapters:

Fictional Visions of Science and Technology

Chapter 2 examines the germinal role fictional visions play in developing a variety of attitudes taken towards science and technology. For the purposes of this chapter, we will examine certain visions commonly associated with but not limited to literary texts. Literary modes of presentation, that is poetical and rhetorical devices often associated with "literature," offer the most accessible means for assessing the dominant and integral role science and technology play in our lives. Philosophical and scientific texts traditionally have provided a dual focus: they evoke anxiety and apprehension in the most sensitive and provocative fashion, as well as images of hope and prosperity through technological innovations.

Selected texts from the writings of Bacon, Hobbes, and Galileo, as well as from the works of Rousseau, Mary Shelley, and Orwell, are used to introduce the linkage between the etymological deliberations suggested in chapter 1 and the fictional labyrinths in and through which visions of truth and history are articulated.

Legacies, Legends, and Enlightenments: The Pretext of Critique

Chapter 3 parallels chapter 2 in that it identifies the historical, philosophical, and textual (linguistic) conditions that make possible the insertion of science and technology into a privileged place in the intellectual world. The contemporary cultural status of science and technology is predicated on a series of fictions—legacies, legends, and enlightenments—that present simultaneously a sense of confidence and suspicion that arise against the background of the scientific revolutions of the fifteenth through the seventeenth centuries and the "Enlightenment."

Instead of identifying a single, unqiue form of Enlightenment, the texts of Kant, Rousseau, and Hume are used to demonstrate a multiplicity of enlightenments predicated on the value assigned to the critiques of reason. The enlightenments articulated in these eighteenth-century texts sustain certain forms of optimism and pessimism that recur in nineteenth-century texts. Further, they sustain the ambiguous interplay of these attitudes regarding the authority of reason's self-critique and the progress of scientific and technological knowledge. The texts of Marx and Nietzsche illustrate explicitly the tenuous nature of enlightenment critique and the diversity of techniques by which such a critique is applied.

The Dissemination of Authority

Chapter 4 presupposes the multiplicity of legacies, legends, and en-lightenments described in chapter 3. Such a multiplicity does not lend itself to a reduction to one source, a single foundation, or a transcendental rule. The multiplicty of enlightenments, for example, is a function of the incessant generation of narratives that declare and legitimate their rule within the discursive labyrinths they constitute.

Classical-canonical and *modern-assessment* views of authority are two traditional ways of appealing to some authority presumed *outside* the narrative networks. The appeal to an authority allegedly *beyond* the mesh of linguistic relations is an issue of discursive replacement: the replacement of one discourse with another thought to be more appropriate in terms of presenting a more comprehensive picture of the world. The universal rules, standards, or criteria presented in these views transcend the critiques that establish them.

The issue of dissemination presents a shift in the analysis of the authority claimed by and attributed to philosophic, scientific, and technological models. Instead of discursive replacement, the dissemination of authority involves discursive displacement, where no first nor last word is ever legitimate. The emphasis is explicitly on linguistic usages. According to discursive displacement, all narratives are "meta-" narratives in that they place themselves at the boundaries of existing discourses. In this way, no narrative can assume the posture of a Meta-narrative or a narrative that lies beyond the labyrinths of narratives.

Consequences of Dissemination: Narrative Recollections and the Languages of Pedagogy

Chapter 5 continues the examination of the dissemination of authority and discursive displacement by introducing and elaborating upon the contemporary-linguistic view of authority. Tracing its articulation in the writings of Rorty, Quine, Kuhn, and Feyerabend, this view of authority attempts to account for discursive displacement in terms of the indetereminacy of translation and the incommensurability of theories. In spite of its focus on a plurality of narratives and interpretive strategies, this view employs the standards of replacement associated with the classical-canonical and modern-assessment views, and finds authority concentrated in certain privileged accounts.

Instead of locating authority in a particular genre or discursive mode, our identification of discursive displacement attempts to show how the fabrication and deployment of rules is pertinent to any interpretive experiment. In order to talk about the dissemination of author-ity, we have used two rules—"use creates" and "all learning is re-collection"—, rules legitimated by their use alone. The use of these rules demonstrates the impossibility of fixing in any permanent fashion the boundaries and limits that constitute cultural matrices. Given this condition, we ask how education can be undertaken except as an ex-periment that is nothing more nor less than an interpretive re-collec-tion of previous experiments?

For the purposes of this project, *language* is no longer conceived of simply as a tool or a set of tools by which we describe a particular state of affairs, "worldly" conditions, or, even, the theoretical and practical interplay of science and technology. Nor is language conceived of merely as a tool for prescribing the conditions of human activity. Based on the writings of Wittgenstein, J.L. Austin, Derrida, Foucault, and Lyo-tard, language is understood in a more productive or performative sense: *language creates the conditions, that is to say the labyrinth of fictions, that make its performance possible; it creates the culture in which it is performed.*[19]

Notes

1. For the purposes of this discussion, the political/social, theological, and scientific technological aspects of contemporary life are incorporated by the reference to the "cul-tural." Very recent discussions of this interlacing have developed from an appreciation and appropriation of Martin Heidegger's and John Dewey's work on technology. See Al-bert Borgmann, *Technology and the Character of Contemporary Life: A Philosophical In-quiry* (Chicago: University of Chicago Press, 1984); Borgmann (with the assistance of Carl Mitcham), "The Question of Heidegger and Technology: A Critical Review of the Litera-ture," *Philosophy Today*, Summer 1987 (Special Issue), 98-191; and Larry Hickman, *John Dewey's Pragmatic Technology* (Bloomington: Indiana University Press, 1990).

2. With the claim that "use creates," we are working with a theme that has been articulated in a variety of ways in recent philosophical, literary, and scientific/technological contexts, i.e., that linguistic usage creates the meaning of a term. For an initial announcement of this motif, see Charles Sanders Peirce, *Collected Papers*, ed. Paul Weiss and Charles Hartshorne (Cambridge: Harvard University Press, 1931-35), 5.400-5.402; Ludwig Wittgenstein, *Philosophical Investigations*, trans. G. E. M. Anscombe (New York: Macmillan, 1968), secs. 43, 138, 197, 454, 556-57; and J. L. Austin, "Performative Utterances," *Philosophical Papers*, 2d ed., ed. by J. O. Urmson and G. J. Warnock (Oxford: Clarendon Press, 1970), 233-52. More recent articulations of this motif from Austin and Wittgenstein are found in: Jacques Derrida, "Signature Event Context," *Glyph* 1 (1977): 172-97; and Jean-François Lyotard, *Le Différend* (Paris: Éditions de Minuit, 1983), pars. 102, 190-92. Langdon Winner has applied this motif to the question "What is technology?" in *Autonomous Technology: Technics-Out-of-Control as a Theme in Polit-ical Thought* (Cambridge: MIT Press, 1977), 9 and 201ff.

3. Gilbert Ryle, *The Concept of Mind* (New York: Barnes and Noble, 1949), 25ff. Ryle maintains the separation of these designations. We refer to "knowing-how" and "know-

ing-that" as convenient terms for designating two sides of a very complex, reciprocal relation between kinds of activity, ways of talking, modes of understanding.

4. The distinction between "scientific technology" and other kinds of technologies, such as political or risk-management, is developed by Joseph Agassi in *Technology: Philosophical and Social Aspects* (Dordrecht: Reidel, 1985), see especially 11-16, 44-51, 152-57, and 191-210.

5. See Plato, *The Republic*, where ruling is discussed as a "skill"; see 347-49 and the parable of the ship 488-89.

6. See Xenophon, *Oeconomicus*, trans. by C. Lord, in Leo Strauss's *Xenophon's Socratic Discourse* (Ithaca: Cornell University Press, 1970).

7. *Encyclopedia Britannica*, 11th ed. (New York: Macmillan, 1911), vol. 26, 497.

8. Peirce, *Collected Papers*, 5.402.

9. Ibid., 5.400.

10. Jacques Ellul, *The Technological Society*, trans. John Wilkinson (New York: Alfred J. Knopf, 1964). See also the most recent account of this issue in Frederick Ferré, *Philosophy of Technology* (Englewood Cliffs: Prentice Hall, 1988).

11. Winner, *Autonomous Technology: Technics-Out-of-Control as a Theme in Political Thought* (Cambridge: MIT Press, 1977).

12. Bernard Gendron, *Technology and The Human Condition* (New York: St. Martin's, 1977); see also Herbert Marcuse, *One-Dimensional Man* (Boston: Beacon, 1964).

13. Godfrey Boyle, David Elliot, and Robin Roy, *The Politics of Technology* (London: Open University Press, 1977).

14. Carl Mitcham and Robert Mackey, ed., *Philosophy and Technology: Readings in the Philosophical Problems of Technology* (New York: Free Press, 1983), 2d ed. See also the following: *Research in Philosophy and Technology*, ed. Paul Durbin (1978-1988) and Frederick Ferré (Westport, Conn: JAI Press, 1988–); *Philosophy of Technology*, ed. Paul Durbin (Dordrecht: Reidel, 1987–); Friedrich Rapp, "Philosophy of Technology," in *Contemporary Philosophy: A New Survey*, vol. 2, ed. Güttorum Floisfod (The Hague: Martinus Nijhoff, 1982), 361-412.

15. Don Ihde, *Existential Technics* (Albany: State University of New York Press, 1983); see also Borgmann, *Technology and the Character of Human Life*.

16. Joan Rothschild, ed., *Machina ex Dea: Feminist Perspectives on Technology* (New York: Pergamon, 1983); also see Joan Rothschild, *Teaching Technology from A Feminist Perspective* (Oxford: Pergamon, 1988).

17. Hans Jonas, *The Imperative of Responsibility: In Search of an Ethics for the Technological Age* (Chicago: University of Chicago Press, 1984). See also, A. Pablo Iannone, ed., *Contemporary Moral Controversies in Technology* (Oxford: Oxford University Press, 1987).

18. Agassi, *Technology: Philosophical and Social Aspects* (Dordrecht: Reidel, 1985).

19. For a recent discussion of this point, from a slightly different perspective and set of concerns, see Ivan Illich and Barry Sanders, *A B C: The Alphabetization of the Popular Mind* (San Francisco: North Point, 1988).

CHAPTER 2
FICTIONAL VISIONS
OF SCIENCE
AND TECHNOLOGY

Introduction

Two interrelated claims have been advanced regarding the interplay of science and technology.

(1) The *activities* incorporated under the headings of *science* and *technology* presuppose and unfold within a broad cultural/ linguistic matrix.

(2) The *meaning* of the terms *science* and *technology*, and thus, by analogy, the *significance* attached to the set of activities designated by these terms, is determined by their *use* within a broad cultural/linguistic matrix.

These claims were advanced in a discussion primarily concerned with the etymology of the terms *science* and *technology*. The appeal to the Greek and Latin roots of these terms showed how *use* creates the means and constitutes the context necessary for announcing the interdependence of science and technology while still preserving their differences. As a result of this appeal, a third claim can be advanced regarding the interplay of science and technology. (3) The knowledge of ideals, principles, and rules, that is to say the knowledge that constitutes theory, is already *use-knowledge*, knowledge arising from the need to satisfy certain practical demands.

The interdependence of theory and practice incorporates and parallels the interplay of the etymological and the textual. The textual can be seen as tracing the domain of practice, while the etymological can be seen as tracing the domain of theory. The interdependence of these domains provides the conditions of reciprocity that inform and thus determine the etymological in terms of textual exigencies and the textual in terms of its etymological (history/use) context.

Moreover, the appeal to etymologies demonstrates how, at once, we work in at least three intertwined domains: the theoretical and practical, the etymological and textual, and the scientific and technological. What must be taken into account, then, is the reciprocity that constitutes these domains. Each domain mediates the others. The etymological/textual mediates the theoretical/practical, and the theoretical/practical mediates the scientific/technological. The reciprocal intervention of the activities and *facts* in one domain with the activities and *facts* in the other domains underlies any attempt to comprehend their assemblage. As such, the *assemblage* of these domains constitutes *the cultural matrix* within which their intersections are recognized.

Fictional devices—whether they are literary texts, poetic images, artistic expressions, philosophical propositions, or scientific theories—are the means by which we approach those points where the domains in question seem to intersect and differ. One way to trace the different applications of the terms *science* and *technology*—that is to say, the *joint articulation* within and between one or more of the identified domains—is to focus on their representation in various fictional accounts. For present purposes, this focus can be achieved through an exploration of specific texts from the literature of the early seventeenth through midtwentieth centuries. The goal of this exploration is to isolate those texts that work with particular, well-defined conceptions of science/technology, and whose fictional presentations or visions are informed by that conception and thus speak to the question of the fictional character of scientific/technological discourse.

To designate any vision of science and technology as "fictional" is to open the field of literature and creative works that attempt to give accounts of the role science and technology play in our lives, both in terms of fact and possibility. Focusing on literary images exclusively would be too limiting. In light of the relatively low level of literacy that existed prior to the present century, certain nonliterary devices, such as folktale and oral history, that played a significant role in the transmission of knowledge from one generation to the next, would be too easily neglected. The place of science and technology in folk medicine or in the proliferation of art in the twentieth century, for example, would be overlooked if the discussion were restricted to the literary world.

To call these visions "poetical" seems somewhat more appropriate in the sense that attention is turned toward the form or style of the discourse or artistic product. Yet, this too limits the scope of the current discussion: one can become bound too easily to the use of the word in its stark verbal or written form. "Fictional," on the other hand, allows for a different kind of discursive openness. It recalls the fabulous nature of representation, while maintaining the literary and the poetical as categories or genres of fictional discourse.

Representation supplies, by way of allusion, the "unpresentable in presentation itself."[1] Representation, or the fictional, then, involves the creation of concepts. A concept is an image that, in its projection, analogous to those images encountered in the cinema,[2] defines its own limits, marks out the boundaries of its field (of objects), and orients the background and foreground of comprehension. As such, fictional accounts or representations localize and isolate possibilities, thus framing and translating the complex interplay among cultural domains into a more condensed field of vision. The representation of science and technology in scientific texts, literary texts, poetry, painting, film, and sculpture follows this pattern. The past or future achievements of scientific and technological investigation, or the potential for catastrophes associated with the unchecked "progress" of science and technology, have an existence only through—are understood only in terms of—what is said, only where they are talked about, represented for a specific purpose. Within a given historical-linguistic context, science and technology are known in terms of their potential, in terms of how their possible achievements or catastrophic consequences are projected, seen in advance, or portrayed in texts of various sorts. Moreover, by focusing on the "fictional," the different uses of the terms *science* and *technology* can be seen as indicating ways by which certain changes have been effected in and absorbed into continued efforts to categorize and demarcate certain forms of knowledge, as well as usher in new world views, or even criticize other perspectives.

In short, in order to present any vision one must presume, as a condition of this envisioning, that the meaning and values assigned to it result from its being put into play, from its contextually bound articulation. Whatever the vision is, it is fabricated or forged for the purpose of presenting a certain, well-defined, well-articulated picture of science and technology. (This depiction, of course, also represents the relationship of science and technology to the so-called liberal arts or humanities.) Any vision is a remaking of the "world" *counter* to what is perceived to be the dominant or governing vision. It is in this sense that the notion of *counterfeiting* underlies all fictional representations.

If representation is the mode or technique for advancing certain claims of truth and knowledge, then these accounts remain fictional as well. Distinguishing between the "fictional" and the "literal," or between the "true" and the "illusory," provides minimal aid in tracking the interplay of science and technology. As a return to specific historical texts will show, these designations are devices for disclosing certain values embedded in specific world views. Therefore, it is necessary to avoid the pretense that a consensus regarding a hierarchy of principles or standards, that is to say a set of criteria against which to evaluate fictional accounts, would bring into relief the lines demarcating the fictional from the true or the factual. These limits or boundaries remain vague; they always remain open possibilities.

Fictions, representations, or the narrative accounts of science, for example, are frames or "posts," as Jean-François Lyotard remarks, "through which various kinds of messages pass."[3] The assemblage of cultural domains appears, at every stage of intersection and divergence, "pluralistic," it appears as a theater in which "several possibilities are immediately [and simultaneously] seen to obtain."[4] Within this framework, meaning and knowledge are not determined nor restricted to a "cognitive core that lies at the heart of a knowable object."[5] Instead, meaning and knowledge, indeed metaphysical and epistemological "foundations," are reestablished incessantly, on their own grounds and in their own terms, at the limits of their use. However, some fictions become "grand narratives,"[6] that is to say great myths or theories that unify the assemblage of domains, smooth over the differences that give this assemblage form, and harmonize the multiplicity of perspectives engendered by the reciprocal intervention of the theoretical/practical, the etymological/textual, and the scientific/technological. Yet, they remain fictional accounts.

Such grand narratives are common within, but certainly not limited to, philosophical and scientific discourse. They attempt to identify those concepts, rules, or principles that validate their own practices, as well as provide standards that will regulate the development of subsequent accounts. A narrative-fiction refers to a particular state of affairs (taken from the past, for instance, but reference to a possible future must be included as well) in order to ground itself, to give itself a legitimate foundation on which to make its claims. In this case, the narrative-fiction is presented *as if* its account settles certain issues associated with or emanating from a state of affairs; *as if*, now, these questions or themes are resolved and completely understood; *as if* certain truths had been decided there already. If use creates meaning, if use creates the means and the context necessary for making *any* announcement possible and meaningful, it is because any fictional account, even when provided under the disguise of a philosophic or scientific treatise, determines the conditions and rules under which it—or science and technology, as the case may be—will function. Fictional accounts, representations, "grand narratives," as Lyotard notes, "determine the criteria of competence and/or illustrate how they are to be applied" in the declaration of those criteria, in the present. "They thus define what has the right to be said and done in the culture in question, and since they are themselves a part of that culture," a product, a fiction of that culture, "they are legitimated by the simple fact that they do what they do."[7]

And yet, despite the attempts to present a unified picture, to articulate the rules that should govern our understanding of the picture, and perhaps because of the attempts to present a comprehensive account of the assemblage of cultural domains as a *whole*, no universal knowledge obtains. Knowledge of the totality as such remains deferred. Fictional accounts provide, at best, fragmented and incomplete pictures of the cultural/linguistic matrix that makes them possible. If, as William James claims, "the world is full of partial stories that run parallel to one another, beginning and ending at odd places," stories that mutually interlace and interfere at points, then "we cannot unify them completely in our minds."[8] Universal knowledge, or a so-called objective standard or foundation, remains a fictive Ideal. It remains another story we tell ourselves about ourselves, our needs, and our goals: another story waiting to be inserted into the practical domain in order to be validated.

For whatever purpose(s) or end(s) an account is given, it is a search for rules, even if it claims to have as its foundation the categories, truths, and rules of texts legitimated by past history. One can assume the legitimacy of familiar categories and rules. But this would be to forget the conditions that make possible those categories and rules—the conditions of use, that is the interplay of theory/practice. The articulation of any fictional account, any representation, or any narrative is at once the formulation of the principles that guide the articulation itself. In other words, the performance in art, the speculation in philosophy, the inquiry in science and technology forge the limits and boundaries of a specific discourse. The search for rules is an implicit acknowledgment that we work without rules from the outset,[9] that the principles that will guide our actions remain in the foreground. In this way, the search for or the determination of categories and rules is a function of and is informed by the reciprocal intervention—the mutual supplementation and mediation—of the domains that constitute the complex matrix we call *culture*.

Recent literature of philosophy, literary criticism, and the discourse of science is replete with textual fictions, narratives, and theoretical accounts guided by the desire to uncover rational—metaphysical, epistemological, ethical, or political—foundations for the critique of science/technology. One such account distinguishes between utopian and dystopian visions of science and technology. On the one hand, the utopian vision, for which the texts of R. Buckminster Fuller[10] and Arthur C. Clarke[11] serve as examples, stresses the great promise contained in scientific and technological innovations. As fictions, utopian texts or visions create a fully developed scenery in which the scarcity of nature is no longer an obstacle for the achievement of humanity's hopes—the attainment of eternal happiness, peace, and the release from the toil, exploitation, and alienation of human existence. Science/technology affirms these hopes and desires through the potential of automated production. Science/technology affirms the promise of economic, political, and artistic freedom.

On the other hand, the dystopian vision, and here one can imagine certain texts by Aldous Huxley,[12] Jacques Ellul,[13] and Eugene Zamiatin[14] as presenting variations on this theme, takes a contrary position. Instead of freedom and happiness, the dystopian stresses the burdens of policing science/technology as well as decrying the potential catastrophes associated with uncontrolled technological and scientific developments. According to the dystopian, the survival of human kind cannot be guaranteed, nor can it be protected by divine intervention. We have been and continue to be careless in our control of science/technology by permitting the creation of conditions that bear our imminent demise. Where the utopian dreams of liberation and peace, the dystopian expresses a fear of enslavement to or destruction by the unwarranted use of scientific/technological innovations.

Like so many similar dichotomies, the juxtaposition of the utopian and the dystopian is a heuristic device. It articulates the conditions, the goals, and the orientation of a specific social or political critique of science and technology. Further, like so many other similar devices or fictional accounts, the utopian/dystopian opposition is incomplete, fragmented, and selective; it is designed for the specific purpose of projecting and protecting certain unexamined—if not unquestionable—beliefs. The dichotomy itself identifies and presents the perimeters of its discourse. Each side of the demarcation—utopian/dystopian—specifies an alignment with or against science/technology. In their respective fashions, each side privileges science/technology. The categories of science and technology perform a specific function within the discourse of the dichotomy. Their contents are already in place; their scope already ordered and comprehensive.

The discussion presented here traces several historical examples representative of specific attitudes taken towards science and technology. In light of the preceding discussion, no text or fictional account could be construed as a comprehensive representation, and the texts engaged in the following discussion are no exceptions. The discussion will traverse literary/philosophical/scientific texts taken from the early seventeenth century, Francis Bacon's *New Organon*, Galileo's *Dialogues Concerning Two New Sciences*, and Thomas Hobbes's *Leviathan*, Jean-Jacques Rousseau's *Emile or On Education* of the mideighteenth century, Mary Shelley's *Frankenstein* of the early nineteenth century, and George Orwell's *1984*[15] of the midtwentieth century. These texts constitute the framework wherein certain themes and questions can be addressed and thus linked to their articulation in other texts and artistic forms—such as twentieth-century atonal music and sculpture, both of which are conceptual functions of science/technology.

One cannot avoid a close examination and exploration of *texts*, fictional accounts, or narratives. Textual narratives supplement the artifacts of artistic expression—paintings, sculpture, film, musical recordings, and the like—which, in turn, supplement narratives. In many instances, textual narratives are the only remains of and thus access to many projections of theory/practice or science/technology. For example, except for numerous paintings, sketches, and sculptures, Leonardo's *Notebooks*, *Treatise on Painting*, and fragments of his literary works outline his planned artistic creations, providing sketches for his scientific/technological speculations.[16] Many of Leonardo's speculations about flight and the principles of aerodynamics, for example, remained unrealized until the development of materials, engine designs, and machinery of the twentieth century translated the images.

Because of its apparent immediacy and so-called timelessness, literature—the written word—remains the primary representation of the diversity and richness of our cultural activities. An appeal to literature or texts has its own virtue: texts work with familiar images, images that are recast for the purpose of reintroducing and rethinking the relationship between science and technology in terms of the reciprocal interplay and intervention of cultural domains presented in this discussion. Moreover, even though specialized (or technical) texts are poor repositories of popular culture and of popular views on science and technology, because they are the fruits of an elite group that writes for itself and responds to the protests of its small membership, these texts always bear the possibility of affecting the lives of those for whom they where never intended. In general, texts become the grounds on which links to a projected set of circumstances or a foreseen future are made.

Some Fictions of Science and the Science of Fiction

A standard view in the history of Western thought classifies Bacon as a philosopher, Galileo as a scientist, and Hobbes as a social/political theorist. However, what has been said about the fictional character of texts in general, and their respective attempts to provide a set of rules, a set of criteria with which to evaluate their own claims, forces a break with this traditional classification. During the seventeenth century there was one fundamental distinction among intellectual disciplines: natural philosophy and moral philosophy. According to this rubric, Bacon and Galileo are "natural philosophers" and Hobbes is a "moral philosopher." Yet, Hobbes is a natural philosopher who debates questions of astronomy and metaphysics with Descartes and others, writing his own "discourse on method"; and further, Bacon and Galileo are moral philosophers insofar as their debates with religious and political authorities are concerned with moral issues, the issues of authority, obligation, law, and individual judgment.

What particular themes or issues organize and orient the texts of Bacon, Galileo, and Hobbes? Each one is interested in the presentation of an ideal world, an abstract mythical world that represents a wholly different set of conditions under which humanity could live. Furthermore, for Bacon, Galileo, and Hobbes, the structure of their respective ideal worlds could be explained rationally. One might imagine that their texts constitute what Jürgen Habermas calls "rational discourse," that is to say a discourse that establishes, by appeal to "rationality" and "consensus," the normative rules that will provide the foundation for the rational communicative action of the commonwealth.[17] Every aspect of their works—the methods of explication, the line of argument, the strategies deployed to deflect possible criticisms and objections—rests on rational grounds. But the solicitation of rationality itself is the method or style of presentation. It results in the establishment of the principles or foundations that legitimate these texts: consensus is achieved through the articulation of the ideal. According to this view, texts propose a reasoned way of critically approaching the world in which they are embedded, presenting an equally reasonable alternative to that set of conditions. In this way, Bacon, Galileo, and Hobbes are precursors of the Enlightenment, the Age of Reason, and can be taken as an appropriate ensemble for present consideration.

But, are they indeed convincing in their proposals? Regardless of their own pronouncements, are these texts not engaged in a rhetoric — identified today as rational or Enlightened — that veils other concerns? According to Michel Foucault, for example, these so-called rational discourses are elaborate disguises that, in one way or another, support the status quo out of which they arise.[18] If these texts convince anyone, they convince the converted, those who already believe the line of thought presented. If the reader is not already a convert, then the rhetoric of presentation is designed to eliminate any sort of critical resistance. The foundation of any discourse is rational insofar as it is instrumental, insofar as it presents an analysis of how to overcome certain present or projected obstacles of nature. The texts under consideration are not repositories of knowledge, scientific or other. They are tools for practice. Each text pursues goals that surpass the articulation of theoretical formulations: to promote or to struggle against the order of certain social and political institutions. And yet, they presuppose the theoretical as the theater in which these debates are to be carried out.

Is the scientific text of the seventeenth century anything other than a kind of fiction in the twentieth century? Is the scientific text of today a figment of the seventeenth-century imagination? The boundaries that demarcate the scientific and the fictional are, as already indicated, sometimes ambiguous and sometimes obscure. Hobbes's fictional examination of certain social, political, and economic issues, as one example, is scientific inasmuch as he adopts a mechanistic framework in order to analyze "the Seat of Power" within the commonwealth.[19] To advance certain claims about the fictional character and status of science is not to say science is inherently meaningless or lacks cultural significance. Nor is it, in any way, to prohibit the comparisons of texts taken from different historical periods. But it is to claim the following: just as the fictional informs and incorporates the scientific, the scientific instructs and converges with the fictional.

Francis Bacon published *The New Organon* in 1620. Though incomplete, the book contains several hundred aphorisms in each of its two parts. Each aphorism can be seen as a representation, the presentation of a particular thematic and conceptual whole. Taken in isolation, each aphorism is analogous to a single frame of a film. Each frame can be analyzed and comprehended; some sense can be made of it. But the frame and the film lack significance, or are of little consequence, without the connection of every frame to the antecedent and subsequent frames. Threads or connecting wires are provided: the aphorisms are numbered and ordered, but their limits remain open, to be set in place by the activities—the interpretations, the applications—that follow. One is advised to read the first aphorism before the second. But can they be read at random? After all, like the filming of a particular series of scenes, Bacon's aphorisms could have been "shot" out of sequence. Is the order in which the aphorisms (or frames) are presented not constitutive of the rules that guide their intelligibility?

The first aphorism of Bacon's *New Organon* reads: "Man, being the servant and interpreter of Nature, can do and understand so much and so much only as he has observed in fact or in thought the course of nature. Beyond this he neither knows anything nor can do anything."[20] From the outset, Bacon sets humanity apart from nature, creating a dichotomy in which humanity remains in servitude towards nature. Servitude is only a first step, we are told later: "Nature to be commanded must be obeyed."[21] Obeying nature means learning how nature operates, what rules and laws guide its course. Once understood in human terms, nature can be controlled, if humanity plays the role of both servant and master.

Understanding nature, according to Bacon, is not an easy task. He counts four sets of "Idols" that obstruct the course of human understanding: "Idols of the Tribe," "Idols of the Cave," "Idols of the Market Place," and "Idols of the Theater."[22] The most important of these idols, for present purposes, are the "Idols of the Market Place," which are described with reference to language: "For it is by discourse that men associate, and words are imposed according to the comprehension of the vulgar. And therefore the ill and unfit choice of words wonderfully obstructs the understanding."[23] Bacon continues his lament and explains why the idols of language—words and names—are the "most troublesome." "For men believe that their reason governs words; but it is also true that words react on the understanding; and this it is that has rendered philosophy and the sciences sophistical and inactive."[24]

Bacon does not differentiate between the sciences and philosophy. For him, discourse itself has become so twisted that it is paralyzed; it is "inactive." In order for language and understanding to become active, practical, and useful, human discourse must transcend the vulgar so it may approach a more sophisticated level of understanding. Empirical observations inform the use and construction of linguistic expressions; they provide the standards by which one examines the stories we tell ourselves about what we understand. These stories, or idols, to use Bacon's term, not only obstruct or obscure our understanding of nature, they inform our understanding as well. However, according to Bacon, idols cannot be trusted because the language of nature precedes that of humanity. But in what language does nature speak? Is there more than one language engaged and encountered in our attempts to understand and control nature?

Assuming with Bacon that there are at least two languages—two modes of expression, two ways of representing phenomena—how is translation possible? An interpretive framework is needed. However, this framework is no longer grounded in an Augustinian theology. The secular move, effected by Bacon, is not antireligious; it *replaces* the divinity of the spiritual word with the divinity of the scientific word. Unlike the divine word, the scientific word can be changed. However, for Bacon the change depends on observation and induction, not speculation and belief in the infallible word of the Church regarding the workings of nature.[25]

Galileo Galilei also insists that certain forms of speculation must be eliminated from the study of astronomy and motion. According to Galileo, the subject matter of speculative philosophy is distinct from the subject matter of empirical inquiry. There are four areas of research in which Galileo demonstrates the significance of the separation: the science of experiments, telescopic astronomy, the principles and laws of motion, and the connection between mathematical expressions and experiential reports.

In his contribution to experimental science, Galileo follows Bacon's advice. If we are to communicate with each other about nature, if we are to understand what nature tells us, we ought to do so on the basis of what our observations tell us—with or without the support of technical devices. Galileo takes Bacon's insistence on the authority of observation one step further by using the telescope as an important technical device to substantiate or refute certain scientific claims regarding nature and celestial movement. As a tool, the telescope has become indispensable for any sort of theoretical, scientific thinking. The technical apparatus—the telescope in this case—mediates the visual field, it transforms the remotely empirical and our understanding of it. Visual perception is no longer limited to the imagination or the comprehension of the natural eye's optical field. The telescope radically extends the field of observation. Galileo and his followers are able to observe celestial objects they have never seen before; objects and planetary relationships about which they have only speculated. The telescope creates the means for "seeing" objects that previously have never been conceived as they appear because of the apparatus. Moreover, the telescope creates the means for new and different kinds of scientific speculations.

Like Bacon, Galileo believes that rational discourse must be grounded in experimental science. Any effort to make sense of nature and our observations of nature, any attempt to argue a particular point of view, relies on experience. The Latin root for *experiment, experiri,* also designates "experience" and "what everyone knows."[26] Our experiences constitute the bases of knowledge. Scientific experiments merely systematize the intimate connection already perceived between human experience (practice) and human knowledge (theory). Science or knowledge, for Galileo, is dependent on technology, and the development of a technological apparatus, like the telescope, is informed always by the general concerns that constitute a theoretical framework.

Based on observation and experiment, made possible by the use of the telescope, and grounded in mathematics, Galileo believes he has proven the Copernican heliocentric theory: the Sun is at the center of the universe and, like the other planets, the Earth rotates around it. But, as I. B. Cohen points out, all of Galileo's conjectures are compatible with Tycho Brahe's system, in which the Earth remains at the center of the universe and the planets rotate around the Sun, which in turn rotates around the Earth.[27] Given such a paradox, one might wonder— on the basis of observation—whether experiments demonstrate what they are intended to demonstrate. Or are there other elements and conditions at work in any proof? Are experiments the sole means by which to achieve true knowledge, as Bacon and Galileo hope? Like the experience or the empirical world they attempt to translate and systematize, experiments are open discourses, subject to reiteration, to reinterpretation, and to retranslation. Their results always remain ambiguous and open-ended: they are informed by so many different theoretical and practical concerns; they emanate from a particular context for specific purposes; and they are used for different purposes.

The *Dialogues Concerning Two New Sciences* [1638] employs four literary or rhetorical devices to demonstrate the need for separating the concerns of empirical inquiry from the concerns of speculative philosophy and theology. The text is constructed as a dialogue that extends over a specific period of time: four days. Each day marks a major thematic division in the text; each day marks the progress of the dialogue. Moreover, the dialogues unfold in three languages: Latin, the universal language of mathematics, science, and theology; the vernacular Italian; and the language of pictorial illustration.

Constructing the text according to these devices, Galileo is free to speak in more than one voice. He speaks through each character, never to be identified with any one. He can be antagonistic, critical, even cynical, without being liable—the *Dialogues* narrate a conversation that takes place elsewhere but is directed to the present. In order to catch the reader's imagination, in order to create some level of identification with the questions and positions taken into consideration, Galileo casts these ideas and arguments not as his own, but as the perspectives of others arguing an issue, debating, and criticizing each other. The dialogue allows anyone to participate. It encourages the open discussion of those contentious points constituting the debate itself, acknowledging the need to involve the nonspecialist as well as the specialist in the exchange. In this way, the dialogue, i.e., language, is the performative medium for determining the legitimacy of any theoretical, philosophical, theological, or scientific perspective. The dialogue stages the debate, it creates the scenes in which the controversy unfolds daily.

Like Bacon's aphorisms, taken as separate units, each day marks a change in temporal and thematic focus. Yet, each day constitutes a link in an infinite chain of "days"—either past or yet to be, either literary or natural. (Even the literary account of the creation, given in Genesis, is recorded according to the events of particular days.) The day in Galileo's *Dialogues* is a fictional device used to frame a variety of topics.[28] But the fiction within the text is not merely a textual fiction. The day symbolizes and reflects the regularity—the orderly, repetitious, diurnal movement—of nature; its revolution and evolution. Past occurrences can be used as guides for the future; night follows day, and day follows night. Each day frames, or captures, the continuous unfolding of nature before humanity. The unfolding never ceases, it does not end with the close of each day. However, each day bears the dawning of the next, enabling the continued inquiry into the workings of nature.

Something different appears with the dawn of each day. For example, at the beginning of the third day Galileo expresses his desire to found a "new science." He is concerned with the power of experiment and "demonstrative reasoning" and not with a new subject matter, motion. "My purpose is to set forth a very new science dealing with a very ancient subject. There is, in nature, perhaps nothing older than motion, concerning which the books written by philosophers are neither few nor small; nevertheless I have discovered by experiment some properties of it which are worth knowing and which have not hitherto been either observed or demonstrated."[29] The new science is not new because motion is new, but rather because the method used to observe and explain it is new. Therefore, he dismisses the speculative discourse of theologians/philosophers regarding motion; they lack a method of demonstration commensurable with their claims.

In its initial format, the mathematical and technical portions of the book were written in Latin, while the physical and more general discussion were written in Italian. Was Galileo assuming Bacon's view on the diversity of languages? Does nature use Latin to express itself, rather than Italian? Or, does the Latin allow for a more accurate translation and universal interpretation of nature? In other words, can nature be comprehended only through one language? Only Latin?

Just as Galileo speaks through the voices of the several characters engaged in the textual dialogue, he employs more than one language to articulate and represent the various issues and questions constituting the debate. In a similar fashion, today we use more than one langauge or jargon in our different roles and occupations to discuss problems that traverse many diverse and heterogeneous domains. On the one hand, one language used today quite commonly is the artificial or formal, so-called universal language of mathematics, logic, or computer science. But even in the domain of artificial language, there are many languages used because each language or idiom provides the means of expression that are defined for particular ends within a specific cultural context. On the other hand, we also use more than one natural language, such as English, French, or German, in which to express ideas, ask questions, and pursue different lines of inquiry. Today, it is not unusual for articles and books, published for both the intellectual academy and the general public, to be written in one tongue while depending on the nuances of another langauge to assist in presenting a line of thought. Is there a need to speak and write in more than one idiom? Some might argue that if an idea, concept, or theoretical stance is understood in English, for example, it does not require the emendations of mathematical formulations or the appeal to etymological roots. Or, one might argue that once a mathematical model is presented, any (nonmathematical) linguistic additions are superfluous.

Galileo recognizes an audience that extends beyond the small circle of European intellectuals and Vatican authorities. There are a variety of readers, some of whom are more competent in mathematics and some of whom are more competent in the vernacular. Unlike Kepler's, Galileo's works are read widely and wield a tremendous influence among the scholars and the nonscholars of his day.[30] Latin was once the universal language of Europe, especially among the educated. But that Galileo's popularity extends well beyond the intellectual elite is due, in part, to the style of his presentation, not only to the importance of the material he presents.

To supplement the use of mathematical formulations, the Latin, and the Italian, Galileo employs another language, or another mode of representation: illustrations, diagrams, graphs, and geometric figures are interlaced with the textual narrative of the dialogue. Like the mutual supplementation of the Latin and the Italian, these images are significant ingredients of the text. They provide another graphic means for expressing and grasping the complex details of the debate. The graphic image appears more immediate—it corresponds more readily to our experiences—than the abstract, symbolic articulation of the "forces of motion."[31] Moreover, in some cases pictorial images complement the written word, making sensibly present the more subtle aspects of nature that science and technology attempt to systematize and articulate. But, is this to claim that pictorial representations are preferable to other languages, as Otto Neurath claims when he says that *"words divide, pictures unite"*?[32] Do pictures unite in the sense of providing a universal language everyone can understand, just as Latin was thought to be in the seventeenth century, and just as mathematics is considered today?[33]

Pictorial representations need not be limited to illustrations that embroider the margins or decorate the cover of a text. They can present the structure of the text itself. Thomas Hobbes fabricates an image of a giant "man" in the *Leviathan* [1651], an image of the commonwealth, the body politic.[34] The commonwealth is, for Hobbes, a product of art not nature. The biblical image of the leviathan is used to capture the artificial construction of social relations as well as the individual's relation to nature. Beginning with the activities of individuals who, in an imaginary "state of nature," are at war with one another, Hobbes traces the conditions that call for the construction of a unified society. Civil society, according to Hobbes, is a necessary, unifying fiction that transforms the relations of strife and struggle to a state of coexistence, in which individuals support one another through a complex set of relations that, in turn, constitute the fabric of civil society.[35]

But nature must be understood too, and individuals must create mental procedures and linguistic devices to guide their comprehension and interaction with it. Once again, in Bacon's *New Organon*, the divinity of the spiritual word is replaced by the divinity of the scientific discourse. In a similar fashion, Hobbes's text replaces the word of God by the word of man or the deeds of science. "Nature (the Art whereby God hath made and governs the World) is by the Art of man, as in many other things, so in this also imitated, that it can make an Artificial

Animal."[36] Hobbes's image is as fabulous as the images deployed by Bacon and Galileo, and yet he too conceives of his work in scientific terms, even though the science in this case (unlike Bacon's and Galileo's) is the "science of natural justice."[37] Since nature is produced by human art, for Hobbes, language is integral to any attempt at systematizing knowledge, especially our understanding of nature.

In *Leviathan*, the "Acquisition of Science" is cast in terms of "the right Definition of Names"; or what, for Hobbes, constitutes "the first" or primary "use of Speech."[38] As he says, "The general use of Speech, is to transferre our Mentall Discourse, into Verbal [discourse] . . . So that the first use of names, is to serve for *Markes*, or *Notes* of remembrance."[39] Speech and language, in general, are the means by which we recall our experiences and articulate our understanding of these experiences. But beyond providing the medium for the systematic expression of experiences, there is a more immediate and integral relation between speech and understanding. "*Understanding*" is, according to Hobbes, "nothing else, but conception caused by Speech,"[40] and speech depends on the connections made to the external world in "Mentall Discourse" (that is to say the definitions and names assigned to natural objects and phenomena). The "connexions" drawn between words, "Names," and the objects they designate are furnished or fabricated through discourse, language—mental or verbal.[41] According to Hobbes's declarations regarding the relation between language, understanding, and nature, how can knowledge claims be evaluated? If "*True* and *False* are attributes of Speech, not of Things,"[42] as Hobbes asserts, then on what objective basis can one claim that any particular representation of nature is preferable to another?

It is in this sense that reason, according to Hobbes, is "not as Sense, and Memory, borne with us; nor gotten by Experience onely; as Prudence is; but attayned by Industry."[43] Experience alone is not sufficient for the development of reason or science. It is human activity, the mental and verbal intercourse with nature and other humans—observation for Bacon and experiment for Galileo—that bridges the natural with the intellectual.

With Bacon and Galileo, Hobbes shares a concern with the role discourse or language plays in the development of science. Bacon recognizes that classical theological claims regarding the language of nature must be supplemented by, if not replaced by, the insights of observation. The languages of Galileo's *Dialogues* demonstrate how the complexities of motion, for example, cannot be rendered intelligible in only one linguistic formulation. Hobbes explicitly examines the definitive function of language in the construction of nature. Science is dependent on discourse that "is governed by desire of Knowledge," but what is commonly called science is "conditionall Knowledge, or Knowledge of the consequences of words."[44] The authority of scientists, as well as the authority of their science, falls short of providing anything more than beliefs: "whatsoever we believe, upon no other reason, then what is drawn from authority of men onely and their writings; whether they be sent from God or not, is Faith in men onely."[45] How can scientific knowledge, understood by Hobbes to be conditional, depend on the authority of scientists and the belief of others in their books? According to Hobbes, knowledge is divided into two categories: "Knowledge of Facts" and "Knowledge of the Consequence of one Affirmation to another." The first sort of knowledge is attained by a witness and is associated with "Facts" perceived through the senses and retained by the memory. It is a kind of knowledge that can be taken as "Absolute." The second sort of knowledge, scientific and conditional knowledge, requires reasoning, intellectual reflection on experience. The books or discourse of science contain "*Demonstrations* of Consequences of one Affirmation, to another." In this respect, they are philosophical treatises, and, therefore, remain suspect in Hobbes's eyes.[46]

Whether it is called philosophical, scientific, or technological, the knowledge amassed and handed down from one generation to the next is an articulation of the stories and myths considered most appropriate for the maintenance of the cultural matrix from which they arise. In other words, the legitimacy and acceptance of a culture, or what may be called a tradition, is constituted in the possibilities of reiteration, in the return to selected texts—stories, myths, fictions.

The texts of Bacon, Galileo, and Hobbes question the univocal authority of the divine word. Moreover, these texts demonstrate how the implementation of a diverse set of linguistic devices and structures is needed for recognizing the multiplicity of ways by which an understanding of nature could be realized. As fictional devices, the discourses and representations of science mediate our experience. But to what extent are these theories—as mediating devices—pedagogically useful?

Science: An Educational Fiction

The image and mediating function of scientific discourse, and the technological devices (sensible, mechanical representations) employed in scientific theories, are given a quite different characterization in certain texts of the eighteenth century. Jean-Jacques Rousseau's *Emile or On Education* [1762], for example, offers an extended criticism of the infatuation with the success of scientific exploration and speculation that marks what we call today the Scientific Revolutions of the seventeenth century.

According to Rousseau, the question facing the education of his fictional student Emile is not "knowing what is but only knowing what is useful."[47] Science presumes the articulation of *what is*: that is to say, science presumes to know nature, to identify the laws that govern natural phenomena, and to explain how we can understand these phenomena. But for Rousseau "the scientific atmosphere kills science."[48] Despite its efforts and claims to capture the essence of what is, science is best characterized by its preoccupation with the development of and dependence on theories and mechanical representations—devices that force withdrawal from nature rather than precipitating closer engagement with nature. From the perspective taken by Rousseau in *Emile*, one can see the efforts of Bacon, Galileo, and Hobbes would have exacerbated the difference between the desire for scientific truth and the spirit of open-ended inquiry that marks Emile's education.

Instead of stressing the theoretical inquiry of epistemology or science, Rousseau says "we are dealing only with practice here."[49] The question that guides Emile's education is "What is it good for?"[50] In other words, what is the practical moment of any (scientific) inquiry? Is it designed to aid in the resolution of a specific problem that a student may encounter? Is it designed to promote and satisfy one's curiosity about what Rousseau calls "the island of humankind" or Earth? Or is it fabricated for the purpose of replacing one speculative account (the theological) with yet another speculative account (the scientific)? Rousseau asks: "What is the use of giving children the idea of an imaginary order which is entirely opposed to the established one they will find and according to which they will have to govern themselves?"[51]

Once set in place, an "imaginary order" or an "intellectual world," one encountered in the Copernican heliocentric model of the universe, mediates our pedestrian understanding of the apparent natural order and the mathematical formulation of stellar rotation. According to Rousseau, naive sensory perception, or Emile's grasp of his environment at this stage of education, is reflected more readily in a Ptolemaic geocentric model of stellar rotation—in which the Earth or, once again, "the island of humankind" is found at the center of the universe. Knowing that the Sun rises in the east and sets in the west is sufficient for Emile's orientation and daily activities. At this point, it is unnecessary for Emile to speculate about the virtues of one theory as opposed to another. There is no necessity in having Emile worry over the Pythagorean mysticism that informs Copernicus's scientific claims, nor the questions regarding how these claims match the spiritual elements of mysticism on any rational grounds.

Even though Rousseau concedes indirectly that the Copernican model may be preferred to the Ptolemaic on wholly theoretical grounds,[52] and that at another stage of Emile's education it may be more useful to Emile than the Ptolemaic commonsense model, Rousseau never relinquishes the belief that it—like any other theoretical apparatus—may be wholly unnecessary. According to Rousseau, it is better that Emile never learns a theory—astronomy or dioptrics—if it cannot be known from experience. " . . . for before he uses these instruments [of science]," Emile's tutor Jean-Jacques claims, "I intend him to invent them."[53]

Adherence to the Copernican model requires an intellectual leap that, for the purposes of educating Emile or even for the purposes of understanding the "natural order" of the heavens on the basis of sensation and experience, results in deception. Emile is neither intellectually nor emotionally prepared to understand and accept the differences entailed by the Copernican model. The Copernican heliocentric model presents a world that is beyond Emile's reach; Emile cannot appreciate the claims to universality and objectivity on which such a theory is predicated. To insist upon the critical *or* naive acceptance of any theory's legitimacy without taking into account the needs of the individuals who may attempt to deploy it is mere pretense. Science that closes off inquiry in the name of the pursuit of truth is deceptive. And science that deceives (intentionally or unintentionally) is to be avoided.

But how is it that the Copernican model or any other scientific theory deceives? How is it that a theory designed to disclose the truths and secrets of the heavens could possibly become a roadblock to inquiry and lead to misjudgment? In order to understand the nature of deception, as it is articulated by Rousseau in *Emile*, it is necessary to rehearse several rudimentary conceptual features of Rousseau's educational text.

Emile is at a point of transition in his education. He is moving beyond being a child of "sensation" to becoming an individual with "ideas," an individual who renders his own judgments. Judgment emanates from simple or complex sensations according to Rousseau, and sensation, which results in either simple or complex ideas, comprises judgment. Ideas are the functions of "the comparison of several successive or simultaneous sensations and the judgment made of them."[54] "In sensation, judgment is purely passive. It affirms that one feels what one feels. In perception or idea, judgment is active. It brings together, compares, and determines relations which the senses do not determine."[55] If theories, in general, provide a specific framework or context for the formation and internal alignment of ideas, then all determination of relations between things in the world will come about according to the prejudices that *already* define the orientation and announce the legitimacy of the theory.[56]

But theories, especially those theories generated in the spirit of science, are content considering only "apparent" or "imaginary" relations. The dependence on mechanical representations and the fiction of theory itself is not conducive to the development of a "precise" and "solid" mind. If Emile is to understand nature, if Emile is to arrive at a point where he forms ideas on "the basis of real relations," "relations such as they are," he must eschew any theoretical apparatus. Why? Because, according to Rousseau, "Nature never deceives us. It is always we who deceive ourselves."[57] And our self-deception comes about through an initial appeal to and subsequent dependence on the presupposition of apparent and imaginary relations proffered in scientific theories: relations that require an intellectual leap that takes one beyond experience, sensation, and judgment.

For Rousseau, it is not sensation, i.e., immediate contact with our environment, that deceives us; our judgments about our environment deceive us. Thus, if "intellectual objects" and "imaginary orders" are to be used, they must be derived from the "objects of sense."[58] The transformation of sensations into ideas, or in other words, the translation of the book of nature into the books of science, does not require a leap of faith into the imaginary world of intellectual objects. Instead, what is required is the use of one's experiences as the practical basis for developing sound and rigorous independent judgment about those experiences and what could be in the future.

Like Bacon, Galileo, and Hobbes, Rousseau appeals to experience as the basis for understanding nature and the laws governing nature. But unlike Bacon, Hobbes, and Galileo, Rousseau's appeal does not presume to offer an objective account of nature and its governing laws. Instead, Rousseau traces a line of thought articulated by René Descartes in *The Discourse on Method* [1636], emphasizing the role of individual judgment in any scientific inquiry. Descartes writes: " . . . no one can so well understand a thing and make it his own when learnt from another as when it is discovered for himself."[59] In a similar fashion, Rousseau claims that "without question, one gets far clearer and

Understood in these terms, the interaction between divine acts of creation and human creative endeavors becomes a continuum that maintains no points of demarcation, only exaggerated sites of convergence and divergence. Moreover, this interaction challenges the pedestrian separation of the Divine and the human. Where Erasmus Darwin's teachings prefigure the Frankenstein fiction as a possibility of medicine and science, they challenge the account of "man" having been created in the image of a Divine Being. And where Charles Darwin's speculations about the origins of the species embrace the possibilities of the transmutation and interrelation of certain species, they too challenge the account of human creation as told in the book of Genesis. If divine "success would terrify the artist," as Shelley asserts, and if "the effect of any human endeavour to mock the stupendous mechanism of the Creator of the world" would be "supremely frightful,"[71] is the only alternative a religious leap of faith, where accepting the textual account of Genesis is imperative? The disquietude accompanying the prospect of humans imitating, duplicating, or replicating divine creation results from the common belief in an inherent gap between the Divine and the mundane. *Frankenstein* challenges this conception by revealing the ambiguity of having to choose between points of origin.

Can one imagine, then, the creation of the human species in terms other than Darwinian evolutionary theory? or in terms other than the account given in Genesis? Could it be imagined that humans could create humans by means other than natural sexual intercourse? Could it be imagined that the reproduction of humans by other means is part and parcel of the evolutionary process?

Like Emile, the young medical student Frankenstein desires to "penetrate the secrets of nature." However, in spite of the many and wondrous insights into nature garnered by modern philosophers, Frankenstein finds himself discontent and unsatisfied with the teachings of science. According to Frankenstein, "the untaught peasant beheld all the elements around him and was acquainted with their practical uses. The most learned philosopher knew little more. He had partially unveiled the face of Nature, but her immortal lineaments were still a wonder and a mystery."[72]

Frankenstein's discontent with science lies in the scientific belief, already announced by Bacon, Galileo, and Hobbes, that nature reveals its truths and discloses its secrets as if they were shells surfaced on the seashore. Frankenstein recalls that "Sir Isaac Newton is said to have avowed that he felt like a child picking up shells beside the great and unexplored ocean of truth."[73] But, as his account of what led him to

science indicates, Frankenstein is not convinced that Newton, nor anyone else, would pick up on nature's truths so readily. Nor is he convinced that one would become confused and lost in the abyss of
nature's disclosures, and thus have to resign oneself to the incompre-
hensibility of nature. Perhaps what troubles Frankenstein about Newton's confidence in science and nature is reflected in Rousseau's observation regarding the inconclusive character of scientific inquiries.
According to Rousseau, the student of science is like a child "on the
shore gathering shells and beginning by loading himself up with them;
then, tempted by those he sees next, he throws some away and picks
up others, until, overwhelmed by their multitude and not knowing
anymore which to choose, he ends by throwing them all away and returning empty-handed."[74]

Again, like Emile, Frankenstein acknowledges the risks involved in
attempting to comprehend the bottomless ocean of truths; in order to
proceed in the face of these risks, nature must be reconstructed in a
manner that differs from the ways it is experienced. That is to say, nature must be fictionalized. To be sure, for Frankenstein nature invites
all students to explore its mysteries, its multiple modes of unfolding.
The most useful avenue for this reconstruction, then, is that of scientific inquiry. But it is a science that comprehends itself as designing and
constituting the world it investigates: a science in which design is not
limited to the production of books, mechanical representations, or
imaginary orders and worlds. Instead, the desire to penetrate the mysteries of nature must be cast as a reconstruction of nature. That reconstruction proceeds by piecing together the world in light of the ingredients washed up on the seashore of the "island of humanity." Where
Robinson Crusoe reconstructs his life along the lines of the civil society
he once knew, Frankenstein attempts to reconstruct life along the lines
of knowledge presumed by scientific society. But in his case, science is
the fabrication, the piecing together of limbs and organs that become
the monster.

Frankenstein's monster, his fiction, is not another device, not an artificial intelligence designed to mirror natural human processes. Instead, in the tradition of Bacon, Galileo, Hobbes, and Rousseau, Shelley demonstrates, through the monster's life, how science fictions or
science as fictions reorient(s) our understandings of and relationships
to nature. Within the context of the fiction Frankenstein, the monster is
real. As a "real" fiction, the monster is a creation of Victor Frankenstein's imagination and laboratory experimentation. Further, as a product of scientific know-how, the monster is brought to life. But its life is

not restricted to the conditions of Frankenstein's imagination nor to the conditions of his laboratory. Constructed in the image of Frankenstein, the monster is free to explore the island of humanity. Unlike the explorations of traditional science, where fictions are employed to make nature speak in a comprehensible langauge, with Shelley the exploration animates the fiction itself. Now the fiction, the monster, has its own voice; now nature, in the form of human creation, speaks and can be understood.

Independent of its creator's hands and wishes, independent of any scientific framework, the fiction eludes comprehension, it defies convention — it has become real. Herein lies its monstrous and sublime presence. The creation has assumed a life of its own, regardless of any of its creator's intentions. Its life is no longer comprehended in terms of fulfilling scientific or divine dreams. Nor are these dreams exhausted in the realization of an image. Dreams become real; but their realization does not necessarily correspond to the reality originally conceived and anticipated. In effect, the monster's presence forces a shift in attention, away from nature as such to the various and differentiated relations we institute with ourselves, nature, and our creations.

How can the voice through which this fiction speaks be heard, if not through the language of its creator? To be sure, Shelley does not resort to a conventional approach to this question. She neither composes another scientific book nor fabricates another tool. Instead, Shelley stages the reconstruction, the recreation, the *identification* of the monster, *as monster*, and its place in a world it has not created. The monster recreates itself in a manner similar to that used by Frankenstein; but now according to a different image. Piecing together it*self*, the monster pieces together the *world*, piecing together *language*.

While hiding next to the cottage of an isolated peasant family, the monster is introduced to the uses of language. As the monster recalls in his own account, "I found these people possessed a method of communicating their experiences and feelings to one another by articulate sounds. I perceived that the *words* they spoke sometimes produced pleasure or pain, smiles or sadness, in the minds and countenances of the hearers. This was indeed a godlike science."[75] The monster's understanding of his surroundings, and the daily activities of the family, is transformed in his attempts to connect sounds to the objects those sounds designate. Moreover, the use of language produces in the monster the very feelings displayed by the individual members of his family. "I cannot describe," the monster exclaims, "the delight I felt when I learned the *ideas* appropriated to each of these sounds and was able to pronounce them."[76]

The fiction completes the work of its creator. Stitching itself together vis-à-vis the lives of others, through the use of language, the monster begins to work with ideas. The possibilities of this specific reconstruction, then, as in the case with any doctoring, are conditioned by words. For the first time the monster entertains an idea of self. As the monster says, "Words induced me to turn towards myself."[77] But that concept of self engendered by language lacks content. Even in Shelley's text, language does not mirror nature. It is only when the monster recognizes that the reflection in a pool of water is, indeed, its own image, that the concept of self is given substance and meaning. The monster "admired the perfect forms" of the cottagers and was "terrified" at the sight of itself "in a transparent pool." At that point, the monster "became fully convinced" that it was "in reality the monster."[78]

The recognition of self, or self-reflection, assumes the presence and activities of other individuals. Moreover, it assumes the possibility of unifying an assemblage of disparate reflections—elements whose origins are unknown. The monster's identity differs from the identity given by its initial design and composition. Instead of seeing itself as the assemblage of the limbs and organs of others, the monster identifies itself with and through the actions and words of others in its own way. Once the monster is capable of such identification, it forces others into self-reflection as well. Thus, prompted by monsters, self-reflection incorporates both a marvel of Divine creation and a warning of what human creation and invention may achieve. This is especially true when the fictions or the products of our creativity take on what seems to be a life of their own, and are not restricted by the uses for which they were once envisioned.

The mystery of nature and the marvel of divine creation no longer dominate the musings of Shelley's text. Shelley marvels at our ability to create the world in which we live: how our understanding of nature restricts our actions and how our fictions define the boundaries of nature and the limits of our actions. Fictions—even the fiction of identity—force us to recognize that the consequences of our activities are not always foreseeable. As such, these fictions—if only in terms of the function they *perform* in any text—warn against the presumption of a fixed end or goal, the presumption of a fixed nature having a specified point of origin, and the presumption of fixed and well-defined contextual boundaries.

Linguistic Construction: Reality/Truth/History

The labyrinth of fictions, already traced in texts as diverse as Bacon's *New Organon*, Rousseau's *Emile*, and Shelley's *Frankenstein*, becomes demonstrably more convoluted in George Orwell's *1984* [1949]. On the one hand, Rousseau and Shelley rely directly on the creation of fictions *within* fictions *about* fictions. Each fiction introduces and reflects a series of possible passages within a maze of paths, where the possibilities are defined in terms of the intertwined pathways that constitute a labyrinth. Within this context, the apprehension of nature *as such* through any form of artifice is suspended.

On the other hand, in spite of their claims to the contrary, the texts of Bacon, Hobbes, and Galileo remain caught within this labyrinth. Even though the ostensible use of fictions in these texts is to provide the means for mirroring nature, the *use* of fictions and narrative devices indicates the difficulty of representing nature in a simple and direct fashion. Regardless of intent or design, the use and deployment of fiction is all that remains in any attempt to reconstruct or understand nature. There is only an infinite play of mirroring within which "nature" appears—a mirroring unable to comprehend the totality of reflections. Thus, no single overview—scientific, political, theological, linguistic, or other—captures the transient characters of reality, history, and truth.

Orwell's *1984* gives an account of the ways by which a dominant and invasive position or "party line" is its own point of reference and, thus, legitimates itself. The image of life in Oceania is one where a complex apparatus is set in place for the purpose of creating fictions that substantiate the authority of the Party. "Newspeak" and "doublethink," the official language and "mental habit" of Oceania, are the two interrelated "technologies" through which the Party's "world-view" is maintained to the exclusion of any other "mode of thought."[79] In Newspeak and the "labyrinthine world of doublethink" this is "reality control."[80]

The power of the Party, the authority of its language and mental habits, is housed in numerous ministries and departments whose function it is to create the fictions of reality, history, and truth. Winston Smith works in the Department of Records within the Ministry of Truth, which "concerned itself with news, entertainment, education and the fine arts."[81] It is Winston's job to rewrite history, "to rectify" earlier, original accounts with the most recent. Discarded historical information or accounts are dropped "into the memory hole to be devoured by the flames." Current accounts, or "speakwritten corrections," are deposited into an "unseen labyrinth" of pneumatic tubes. Controlling the reconstruction of history through the bits and pieces of information that appear within the partitions of his cubicle, Winston is an accomplice in substantiating the invincibility of the Party.

It is in this sense, as Winston's position indicates, that there is no point of origin, there is no past; there is only the control of what is now, what is "still happening" — that is, the future. As such there are no copies, there are no falsifications. As such, there is no deception. Everything that is "true now was true from everlasting to everlasting."[82] As Orwell notes, for the Party, "all history was a palimpsest, scraped clean and reinscribed exactly as often as was necessary."[83] There is no connection to a reality against which the reconstruction can be measured. The real is fictional, but no less real. The real is counterfeit: it is a substitute for that which is no longer true, whether it is accounted for in the Department of Records or another department within the Ministry of Truth, the Fiction Department. As Winston comes to realize, any piece of information, even statistical data, and the means used announce it are "just as much a fantasy in their original version as in their rectified version."[84]

Newspeak and doublethink — the authority of any means for determining what is real — are substitutes for the discourse of scientific thought. Yet, they too require the technologies of disciplined thought. And where discipline does not come naturally, some form of education, if not the training of disciples, becomes inescapable. As O'Brien, an inner-Party official, reminds Winston, "Only the disciplined mind can see reality. . . . You believe that reality is something objective, external, existing in its own right. You also believe that the nature of reality is self-evident."[85] Where formal education fails to create the proper mental habits, and where being a disciple is unpalatable, only the extreme measures of technical control, i.e., physical, intellectual, and emotional torture, can bring about a disciplined world view. As far as O'Brien is concerned, "it is impossible to see reality except by looking through the eyes of the Party."[86]

The absence of reference to anything beyond the symbols used in Newspeak and doublethink guarantees their authority. In other words, the Party is founded on the presumption that "reality is inside the skull. . . . Nothing exists except through human consciousness."[87] Unlike the commonly held world view of the letters, arts, and sciences of the seventeenth, eighteenth, and nineteenth centuries, where consciousness extends itself towards reality to control it, with Orwell there is no longer any bifurcation of the real and the fictional. If reality is, indeed, inside the skull as O'Brien claims, then only the Party, or what is tantamount to the adherence to any "story" or "technology of fiction," can establish the laws governing society and the laws governing nature.

The eventual move from the astronomy of Ptolemy to the astronomy of Galileo involves the displacement of one series of fictions by another series. Every text carries the possibilities of displacement. In *1984*, Orwell recognizes that displacement is not identical to replacement. As O'Brien asks Winston, "Do you suppose it is beyond us to produce a dual system of astronomy?" After all, in contemporary science Newtonian mechanics, Einsteinian physics, and quantum mechanics are used simultaneously. Given this situation, O'Brien responds to his own question by saying, "The stars can be near or distant, according as we need them."[88]

Replacement occurs only when memory no longer serves to recollect any other series of fictions that have ever been composed or encountered. In displacement any fiction can be used again, for purposes other than those for which it may have been designed. Displacement, then, engenders repetition and thus thwarts binary opposition. Under these conditions, can one imagine anything outside a labyrinth of fictions, that is, outside the text? Or is all imagining bound to this labyrinth?

Notes

1. Jean-François Lyotard, *The Postmodern Condition: A Report on Knowledge*, trans. by Geoff Bennington and Brian Massumi (Minneapolis: University of Minnesota Press, 1984), 81.

2. See Gilles Deleuze, *Cinema I: L'Image-Mouvement* (Paris: Les Éditions Minuit, 1983), 46-61 and 243-289; see the English translation by Hugh Tomlinson and Barbara Habberjam, *Cinema I: The Movement-Image* (Minneapolis: University of Minnesota Press, 1986), 29-40 and 178-215.

3. Lyotard, *The Postmodern Condition*, 15.

4. William James, "The Moral Philosopher and The Moral Life," in *The Will to Believe and Other Essays in Popular Philosophy* (New York: Dover, 1956), 191.

5. Michel Foucault, "Theatrum Philosophicum," *Language, Counter-Memory, Practice*, ed. by Donald F. Bouchard (Ithaca: Cornell University Press, 1977), 174.

6. See Lyotard, *The Postmodern Condition*, 7-9, 18-23, and 31-41.

7. Ibid., 23.

8. William James, *Pragmatism* and *The Meaning of Truth* (Cambridge: Harvard University Press, 1978), 71.

9. See Lyotard, *The Postmodern Condition*, 81ff.

10. R. Buckminster Fuller's utopian accounts are too numerous to list. But, for present purposes, see *Earth, Inc.* (Garden City, New York: Doubleday, 1973); *Utopia or Oblivion: Prospects for Humanity* (New York: Bantam, 1969); *Synergetics*, Vol. 1 and 2 (New York: Macmillan, 1975 and 1982); *Critical Path* (New York: St. Martin's, 1981).

11. There are many accounts of the future presented by Arthur C. Clarke. But see especially, *Profiles of the Future* (New York: Dell, 1968); *2001: A Space Odyssey* (New

York: Dell, 1969); and *July 20th, 2019: Life in the 21st Century* (New York: Macmillan, 1986).

12. Aldous Huxley, *Brave New World* (New York: Harper & Row, 1969).

13. Ellul, *Technological Society*.

14. Eugene Zamiatin, *We*, trans. Gregory Zilboorg (New York: Dutton, 1952).

15. George Orwell, *1984* (New York: New American Library, 1981).

16. See *Notebooks of Leonard da Vinci*, ed. Irma A. Richter (Oxford: Oxford University Press, 1982); *Treatise on Painting by Leonardo da Vinci*, ed. A. P. McMahon (Princeton: Princeton University Press, 1956); and *The Literary Works of Leonardo Da Vinci*, ed. A. P. Richter (London: Phaidon, 1970).

17. Jürgen Habermas, *The Theory of Communicative Action*, vol. 1: *Reason and the Rationalization of Society*, trans. Thomas McCarthy (Boston: Beacon, 1981); see esp. ch. 1.

18. Michel Foucault, *The Order of Things: An Archeology of the Human Sciences* [1966] (New York: Vintage Books, 1973). See especially chs. 2, 3, and 5 of Pt. 1, and chs. 7 and 9 of part 2.

19. Thomas Hobbes, *Leviathan, or the Matter, Forme, & Power of a Common-Wealth Ecclesiasticall and Civill* [1651], ed. C. B. McPherson (New York: Penguin Books, 1968), 75.

20. Francis Bacon, *The New Organon* [1620], ed. Fulton H. Anderson, New York: Macmillan, 1985, 39.

21. Ibid., Aphorism 3, 39.

22. Ibid., Aphorism 39, 48.

23. Ibid., Aphorism 43, 49.

24. Ibid., Aphorism 59, 56.

25. Bacon's methodological recommendations are classified in the philosophy of science as "naive inductivism." According to this view, one ought to assemble as many observations as possible in order to formulate some general statement, a law of nature. Once the general statement is formulated, it can be revised continuously in light of new observations, either confirming or refuting it. More sophisticated accounts of induction include an adherence to the theory of probability and acknowledge the difficulty of comparing empirical observations with statements *about* empirical observations.

One recent exploration of Bacon's philosophy of science is Peter Urbach's *Francis Bacon's Philosophy of Science: An Account and a Reappraisal* (La Salle, Ill.: Open Court, 1987). Urbach explains the complexity of Bacon's methodology in comparison to that of Karl Popper, while remaining a defender of Bacon's intent.

26. I.B. Cohen, *Revolution in Science* (Cambridge, Harvard University Press, 1985), 140.

27. Ibid., 136.

28. See Galileo Galilei, *Dialogues Concerning the Two New Sciences* [1638], trans. Henry Crew and Alfonso de Salvio (New York: Dover, 1954), "First Day," 6-11. Here the characters of the dialogue debate the methodological role thematic digressions play in the attempt to provide a specific scientific demonstration. Even though the dialogue sets out to avoid philosophical speculations of the Augustinian sort, "theoretical speculation" remains part and parcel of scientific discourse. In fact, Sagredo asks, "If, by digressions, we can reach new truths, what harm is there in making one now, so that we may not lose this knowledge . . . remembering also that we are not tied down to a fixed and brief method but that we meet solely for our own entertainment? Indeed, who knows but that we may thus frequently discover something more interesting and beautiful than the solution originally sought?" (7-8).

29. Ibid., 153.

30. Ibid., 135.

31. See Ibid. 114.

32. Otto Neurath, *Empiricism and Sociology*, ed. Marie Neurath and Robert S. Cohen (Dordrecht: Reidel, 1973), 217.

33. Neurath's view of the scientific foundation of pictorial representations is linked to his acceptance of epistemological positivism. Opposed to this view, one can turn to Nelson Goodman, for example, who claims that pictorial representations are themselves open to interpretation, according to the rules of convention. See Nelson Goodman, *Languages of Art: An Approach to a Theory of Symbols* (Indianapolis: Hackett, 1976); *Ways of Worldmaking* (Indianapolis: Hackett, 1978); *Fact, Fiction, and Forecast*, 4th ed. (Cambridge: Harvard University Press, 1983).

34. In this context we should recall Plato's *Republic*. The *Republic* presents an "image" of the state which, in turn, supplies the format for social criticism and the promotion of certain political issues pertinent to ancient Greek society. See also, Langdon Winner, "*Techné* and *Politeia*," in *The Whale and the Reactor* (Chicago: University of Chicago Press, 1986), 40-58, for a discussion of "political *techné*."

35. Hobbes, *Leviathan, or the Matter, Forme, & Power of a Common-Wealth Ecclesiasticall and Civill*, 81-82.

36. Ibid., 81.

37. Hobbes's own insistence that his work is "scientific" is stressed in the titles (and content) of some commentaries on his work. Compare M. M. Goldsmith, *Hobbes's Science of Politics* (New York: Columbia University Press, 1966); Thomas A. Spragens, Jr., *The Politics of Motion: The World of Thomas Hobbes* (Lexington, Ky.: The University Press of Kentucky, 1973).

38. Hobbes, *Leviathan*, 106.

39. Ibid., 101.

40. Ibid., 109.

41. Ibid., 102: "The manner how Speech serveth to the remembrance of the consequence of causes and effects, consisteth in the imposing of *Names*, and the *Connexion* of them."

42. Ibid., 105.

43. Ibid., 115.

44. Ibid., 130-131.

45. Ibid., 134.

46. Ibid., 147-148.

47. Jean-Jacques Rousseau, *Emile or On Education*, trans. Allan Bloom (New York: Basic Books, 1979), 166.

48. Ibid., 176.

49. Ibid., 81.

50. Ibid., 179.

51. Ibid., 186.

52. Ibid., 168-69.

53. Ibid., 206. Earlier in bk. 3, Rousseau writes that Emile shall never know the philosophical ceremony of distinguishing between "appearance" and "reality." For Emile there is "no appearance and always reality . . . Let him always produce his masterpiece and never pass for a master; he should prove himself a worker not by his title but by his work. . . . He must work like a peasant and think like a philosopher so as not to be as lazy as a savage. The great secret of education is to make the exercises of the body and those of the mind always serve as relaxations from one another" (202).

54. Ibid., 203.

55. Ibid.
56. Ibid.
57. Ibid. "Our greatest ills come to us from ourselves" (48).
58. Ibid., 168.
59. René Descartes, *The Discourse on Method of Rightly Conducting the Reason and Seeking for Truth in the Sciences*, in *The Philosophical Works of Descartes*, vol. 1, trans. Elizabeth S. Haldane and G. T. R. Ross (Cambridge: Cambridge University Press, 1981), 124.
60. Rousseau, *Emile or On Education*, 176.
61. Ibid. Rousseau writes in the subsequent paragraph: "You want to teach geography to this child, and you go and get globes, cosmic spheres, and maps for him. So many devices! Why all these representations? Why do you not begin by showing him the object itself, so that he will at least know what you are talking to him about?"
62. Ibid., 184. Earlier in *Emile*, bk. 2, Rousseau explains what he takes to be the only acceptable function of books. "He [Emile] must know how to read when reading is useful to him; up to then it is only good for boring him." (116)
63. Ibid., 184-88.
64. Ibid., 185.
65. Daniel Defoe, *Robinson Crusoe* (New York: Bantam Books, 1981), 61.
66. Ibid., 59-60.
67. Charles Dickens, *Hard Times* (New York: Bantam Books, 1981), 26.
68. Ibid., 44-45.
69. The subtitle to Langdon Winner's *Autonomous Technology, Technics-out-of-Control as a Theme in Political Thought*, provides us with a useful metaphor for recognizing the appearance of a recurring theme about the unanticipated possibilities of developing technology.
70. Mary Shelley, *Frankenstein* (New York: Bantam Books, 1981), xxiv.
71. Ibid., xxv.
72. Ibid., 25.
73. Ibid.
74. Rousseau, *Emile or On Education*, 171-72.
75. Shelley, *Frankenstein*, 96-97. Emphasis added.
76. Ibid., 97. Emphasis added.
77. Ibid., 104.
78. Ibid., 98.
79. Orwell, *1984*, 246.
80. Ibid., 32.
81. Ibid., 8.
82. Ibid., 32.
83. Ibid., 36.
84. Ibid., 37.
85. Ibid., 205.
86. Ibid.
87. Ibid., 218.
88. Ibid., 219.

CHAPTER 3
LEGACIES, LEGENDS, AND ENLIGHTENMENTS
The Pretext of Critique

Introduction

Negotiating the intricate assembly of any labyrinth (or text) requires dismantling and abandoning, or at the very least suspending, certain classical categories and binary oppositions. But such a dismantling or suspension of concepts is possible only as a reinscription of those concepts, or as a reinterpretation of the text. The textual interpretations of Bacon, Hobbes, Galileo, Rousseau, Shelley, and Orwell, offered in the preceding chapter, demonstrate how the play of certain categories, e.g., objective/subjective, true/false, real/fictional, literal/figurative, and inside/outside, constitutes, in each case, a labyrinth of textual conditions that prefigures a certain vision or account of science and technology. Moreover, the encounter with these texts demonstrates how the boundaries of each category are blurred at the outset, and subsequently set in place to achieve a specific end within a certain context.

In one respect, to describe a labyrinth as such is to present the image of a unity: a clearly defined (or definable), comprehensible whole. But a labyrinth of fictions and narrative accounts cannot be prescinded from the elements and conditions that constitute it. A labyrinth becomes one more fiction or image among a multiplicity that establishes it and links it to an assemblage of other textual fictions. It is an incessant generation of narrative-fictions, then, that makes it possible to speak of a labyrinth. But what, if anything, conditions or makes possible this generation? For the moment, it seems to be the simultaneous possibilities of displacement (fracturing boundaries) and assembly (creating an ensemble). The generation of narrative-fictions presupposes, at once, the displacement of one narrative by another, that is the comprehensive character of displacement, and the possibility of transforming this process of displacement into a totality.

Within the history of philosophy, it is not uncommon to transform the many intellectual and artistic trends associated with a specific historic epoch into a common movement that traverses the period, and thus represents a way of thinking unique to that period. Any intellectual epoch can be identified by reducing its traditions and legacies to a particular tradition and legacy—as if there is a *consensus* regarding the significance of the moment, and as if this significance resides in the thought of certain texts. The *Enlightenment* signifies a definite period in which the pretensions of religious and political authority are betrayed through the critical function of reason. In a provisional manner, it can be said that however reason is pictured—as a faculty or simply as a concept—by the various texts and trends of the eighteenth century, it is linked to the scientific revolutions and technological innovations of the sixteenth and seventeenth centuries. In a similar vein, though not directly related to the present discussion, the *Renaissance* is treated as if it encompasses the rebirth of independent thought in accordance with the revival of a specific Greek world view and thus a specific set of ideals.

Today, the spirit of Enlightenment inquiry pervades the context of science and technology studies. To speak in terms of fictional or narrative labyrinths (or texts) is one way of rethinking the prominent themes and issues associated with the Enlightenment. Just as reference can be made to an elaborate and complex interplay of fictions constituting scientific or literary texts, by an analogous structure, reference can be made to an assemblage of *enlightenments* rather than *the* Enlightenment. That is to say, to speak of the Enlightenment is a convenient fiction used to simplify the diverse texts and attitudes that characterize a specific historical-philosophical era. Within the eighteenth century, then, one can identify complementary and contradictory texts and images that either claim to represent the spirit/essence of (the) Enlightenment or pursue the prospects of different enlightenments. In order to present the differences these possibilities indicate (especially regarding an analysis of science and technology), certain textual themes from the writings of Immanuel Kant, Jean-Jacques Rousseau, and David Hume, will be engaged, in juxtaposition to one another, to exemplify a labyrinth of enlightenments that frame what is today called the Enlightenment.

To speak of enlightenments, then, is to recognize more than one legacy; to speak of enlightenments is to recognize the interlacing of legacies. The legacies, that is those principles of thought and action on which the Enlightenment spirit stands, do not appear as responses independent of the interplay of enlightenments, or what has been called a labyrinth of narrative-fictions. They are given already, although not always clearly articulated, in the juxtaposition of and within the texts under consideration. In other words, these texts, as is the case with any text, create the conditions of their own betrayal. Thus, it seems fitting to disengage the reference to one legacy or another as it pertains to the Enlightenment and subsequent epochs in the history of philosophy. As will be seen, the texts of Karl Marx and Friedrich Nietzsche appropriate certain ideas from more than one legacy. They voice the hope—pronounced in the spirit of optimism—and the despair—pronounced in the spirit of pessimism—concerning the dependence of contemporary culture on the progress of science and technology. Marx and Nietzsche maintain an ambiguity that bears traces of the ambivalence toward science and technology—the testing of reason through logic and experimentation—already articulated in the texts of Kant, Rousseau, and Hume. It is not an ambiguity, an undecidability, that appears after the fact of there having been the Enlightenment. It is an ambiguity that defines enlightenment.

As indicated above, the Enlightenment is the peg, the leg (hence, the legacy) that grounds contemporary discussions regarding the authority and performance of science and technology. The Enlightenment bequeaths its own legend. As a legend itself, the Enlightenment is housed in the stories it presents about the nature of nature and the nature of human beings. Further, as a legend, the Enlightenment lives on in stories told about its force and significance; its spirit and principles are commemorated in the repetition of its narratives and so-called fundamental themes. Through its reiteration, the Enlightenment appears to some as an unbroken project. Though the foundations are set in the eighteenth century, to some the project's completion remains an ideal that must be realized.

The legend that is the Enlightenment focuses on the acquisition of knowledge. Even though individual liberties and the universal emancipation from the authority and oppression of religious dogma, political rule, and speculative doctrines are emphasized by the culture of Enlightenment, there is no question regarding the possiblity of knowledge. In general, knowledge is possible. Just as the texts of Bacon,

Hobbes, and Galileo announce a belief in the ability of science to guarantee knowledge of nature and human beings, what is called the Enlightenment is marked by the general belief in the capacities of reason to grasp the intricacies of physical nature and the secrets of human nature. Moreover, there is a pervasive belief that anyone can master nature by mastering reason. Nature can be known, understood; and human beings can be comprehended in terms of how they act as individuals constituting a community. The means for achieving certainty is reason, the faculty of logical connections, critical evaluation, experimental testing, and judgment. Nature is there: it is comprehensible; it requires no Divine revelation.

One of the most dramatic and intricate expressions of faith in the power of rationality, and in the ideals of the Enlightenment, is found in the *The Declaration of Independence* of the United States of America [1776]. As a declaration, this document is an appeal to reason, an appeal to a universal concept: all humanity, the "candid world," can understand the plight of the colonies once the facts have been presented. The facts include the specific grievances against the British crown and the list of specific individual rights and liberties. These facts are constructed and presented as the premises of an argument. The declaration of independence itself, then, is the only conclusion that can be inferred; it is the only course of action that logic—political or philosophical—dictates.

Beyond the appeal to reason as the faculty or instrument of argumentation, the *Declaration* appeals to reason as the faculty that comprehends the laws of nature. The text, the narrative that is the declaration, conditions its own authority; it is the foundation for independence. In other words, there is no appeal to any source of authority outside the text itself, outside the act of declaration—in effect, outside language. Thus, the laws of nature, to which the text turns for support, are textual laws. As is the case with many seventeenth-century narratives—again, the texts of Bacon, Hobbes, and Galileo are examples—, nature is a book, a text to be deciphered and comprehended. Not only does the text legitimate the laws it claims are "natural," it legitimates the "truths" it cites as "self-evident." Natural laws and self-evident truths can be understood by everyone, regardless of political orientation or religious commitment. And yet, the entitlement of an individual's "inalienable" rights is grounded in the appeal to Divine Providence. Just as the declaration announces and thus creates the independence of the colonies, it declares the ultimate source of its own authority—"Nature's God."[1] The ostensible appeal to an authority beyond the internal structure and thematic alignment of the text, or beyond social conventions, is itself sanctioned by nothing other than the text and its labyrinth of laws and truths, its labyrinth of fictions. In other words, the declaration substantiates the authority attributed to Divine Providence and its role in the conduct of human affairs.

The power, the force, and the significance of the *Declaration* is situated and played out within (the labyrinth of) its own narrative. The drama of revolt—the performance of declaring independence—is and remains textual. But to advance such an assertion is not to claim that the birth of a nation can be reduced to the articulation of specific principles in a set of documents. Indeed, the *Declaration* indicates the contrary. The simultaneous appeal to reason, the laws of nature, self-evident truths, and Divine Providence is the deployment of fictions forged as objective, rational principles.

The fictions are deployed—simultaneously in many different ways—in order to create another fiction, that is, to achieve consensus when it is absent. Different rules and principles are being forged, but they are not yet in place. In order for these rules to have any legitimacy, (the fiction) consensus must accompany their institution. The authority and significance of the *Declaration*, as is the case with any other text, lie in the faith its fictions command subsequently. The fictions, that intricate narrative constituting the *Declaration*, presuppose and unfold on the pretense that a qualitatively different order or system can be established. In an analogous fashion, the authority and significance of the Enlightenment are underwrittten by the belief in its legacies and legends of reason.

Three Critiques of Reason

In the *Critique of Pure Reason*, Immanuel Kant claims that "Our age is, in especial degree, the age of criticism, and to criticism everything must submit."[2] One can use Kant's assertion as a point of reference for understanding the configuration of displacement taking shape in the eighteenth century. Indeed, Kant's statement seems to express the spirit of many ideals and principles associated with the Enlightenment. The texts of Bacon, Galileo, and Hobbes have been used to depict the shift from medieval obedience to theological dogma and faith in speculative systems to a more pronounced faith in the power of reason and the workings of science. Kant identifies with this legacy by focusing on the *critical function* of reason. For him, the "age of criticism" is characterized by "a call to reason to undertake anew the most difficult of all its tasks, namely, that of self-knowledge. . . . "[3] The point of this calling is enlightenment. Thus, for Kant, as for Rousseau and Hume and many others of that period, such as the French *philosophes*, the age of criticism is the age of Enlightenment.

Kant's legacy in the history of philosophy is marked by his three critiques: *Critique of Pure Reason* [1781/1787], *Critique of Practical Reason* [1788], and *Critique of Judgment* [1790]. However, Kant's critiques are designed to issue only a specific enlightenment, one grounded in the universal laws of critical reason. If enlightenment is as closely linked to criticism, especially the critique of reason, as Kant announces, and if criticism assumes a multiplicity of forms or disguises, then not only are there "critiques" of reason but there are "enlightenments" about which one can speak. The texts of Jean-Jacques Rousseau and David Hume provide the links to these *other* enlightenments.

Like Kant, Rousseau and Hume, in their respective ways, are concerned with critiques of reason (or understanding or imagination or judgment) as they pertain to knowledge of the self, nature, and the conditions for constituting the community. But unlike Kant, Rousseau and Hume, again in their respective ways, differ in how they analyze reason, in what they understand the capacities of reason to be in conducting a critique, and what can be accomplished by a critique. Moreover, they question to what extent reason must be faithful to a specific set of values and ideals in order to perform its critiques. So if there are at least three strategies for conducting a critique of reason, as articulated by Kant, Rousseau, and Hume, then are there not three different enlightenments—legacies and legends—that inform the treatment of science and technology from the eighteenth century to the present?

Kant: The Optimism of Enlightenment

Kant's critique of reason is designed to determine the limits of reason, the boundaries of philosophical discourse, and the perimeters of scientific inquiry. The "chief question" he poses in the *Critique of Pure Reason* is this: "How much can the understanding and reason know apart from all experience?"[4] What are the conditions that make knowledge—of nature and self—possible? In an early section of the *Critique* Kant provides a provisional response. "Reason, holding in one hand its principles, according to which alone concordant appearances can be admitted as equivalent to laws, and in the other hand the experiment which it has devised in conformity with these principles, must approach nature in order to be taught by it."[5] Reason's approach to and apprehension of nature is conditioned by the synthesis of two

fundamental sources: universal a priori concepts of pure understanding and intuitions of sensible perception. As Kant's famous dictum
states, "Thoughts without content are empty, intuitions without concepts are blind."[6]

According to Kant, knowledge is impossible without the union of
both faculties. The faculties of understanding and sensibility bring specific powers and conditions to the creation of knowledge. On the one
hand, it is through sensible intuition that an object is given; sensible
intuition provides the receptive faculty for representing objects of
knowledge. On the other hand, an object represented as such is made
intelligble, or is thought, when brought under the universal rule of the
categories or concepts of pure understanding. The capacity to represent or reproduce an object of nature presupposes the universal conditions of space and time. Every object must be presented (extended)
in space and time. Moreover, the capacity to represent presupposes
that the representations are subject to the rule of the categories. The
possibility of *unity* — the possibility of knowing any object of nature as
a manifold of sensibility—is given a priori as one of the categories
found in Kant's Table of Categories under the heading of *Quantity*.[7] To
distinguish a forest *as such* from a forest as an aggregate of trees and
grasslands depends on the a priori deployment of two different categories (or concepts), those of unity and plurality.

Regardless of the conditions limiting the functions and powers of
human reason, Kant believes that within these perimeters, science —
the systematic configuration of *all* knowledge, according to the critique of pure reason—is possible.[8] This is a theme Kant uses to guide
the critique. Towards the end of the *Critique of Pure Reason* he explains, in different terms, that science is possible due to the spontane-
ous systematizing functions of reason. Reason is "by its nature
architectonic."[9] Kant continues:

> The critique of pure reason, in the end, necessarily leads to
> scientific knowledge . . . . When once reason has learnt
> completely to understand its own power in respect of objects
> which can be presented to it in experience, it should easily
> be able to determine, with completeness and certainty, the
> extent and the limits of its attempted employment beyond
> the bounds of all experience.[10]

To say that reason orders nature is to say that nature cannot be apprehended without conforming to the rules that constitute the framework of reason. Therefore, nature cannot teach reason anything reason does not itself order. The possibility of nature—as a totality, a

plurality, or a unity—lies a priori in the understanding, that is to say exclusively within the powers of reason. In Kant's words, "[understanding] is itself the lawgiver of nature. Save through it, nature, that is, synthetic unity of the manifold of appearances according to rules, would not exist at all. . . ."[11]

The order reason imposes on nature is universal. Understanding is, for Kant, "*the faculty of rules*" or laws, where the concepts of pure reason are the rules.[12] Where sensible intuition provides the representation of possible objects, or what Kant calls the "forms" of intuition, understanding provides the universal rules or laws that determine the configuration and content of these forms.[13] The application of concepts or categories of pure understanding to the representations of sensible intuition is, then, necessary in every instance of possible knowledge. It is an application that takes place in an objective and uniform fashion.[14]

From the outset, the organizing principle of Kant's critique is to institute a set of first principles for philosophy and science based on the critical use of reason, as opposed to perpetuating speculative systems of metaphysics that result from the unregulated use of reason. The task of critical metaphysics is to articulate the transcendental conditions of knowledge, that is those fundamental conditions that make any and all knowledge possible. It is not the task of critical metaphysics to posit the existence of "objects" on the basis of "ideas" or concepts. As Kant notes, "transcendental philosophy is only the idea of a science, for which the critique of pure reason has to lay down the complete architectonic plan."[15] To realize the idea of a science that is universally applicable—indeed, a science that forms the ground for all other empirical and practical inquiry—is to present a system that articulates, in a comprehensive fashion, "all principles of pure reason."[16]

Categorizing all principles of pure reason is not limited to Kant's metaphysical discourse; it extends to the discourse on ethics and morality (the critique of practical reason) and the discourse on beauty and the sublime (the critique of practical judgment). The possibility of a coherent system of metaphysics—one that is universal, one that comprehends the discourses of all three critiques—presupposes what Kant terms the "universalizability" of "subjective maxims." In effect, Kant's critique of reason is performed in accordance with this general principle. Recognizing the maxims that govern its subjective performance, reason establishes those laws that can become universal. The move from the subjective to the universal constitutes a reflective turn that enables reason to undertake the task of self-knowledge. "To make use of one's own reason," writes Kant, "means nothing more than to ask one's self, with regard to everything that is to be assumed, whether he finds it practicable to make the ground of the assumption or the rule which follows from the assumption a universal principle of the use of his reason."[17]

In order for reason to perform its self-critique, it must guarantee its own freedom. In other words, reason must legislate the universal empirical use of its principles. The significance of this move, then, lies in the authority it grants reason. By analogy, the individual assumes the authority to impose one's own laws on oneself to the extent that these laws can be applied to others. Kant embraces the ideal of (natural) laws governing both nature and society. However, if reason, or an individual, is to be bound by any law, Kant prefers to have the law(s) self-legislated. Reason or individual will should never bow "under the yoke of laws which others impose on it."[18] According to Kant, "Tutelage is man's inability to make use of his understanding without direction from another." However, "Enlightenment is man's release from his self-incurred tutelage."[19] The freedom from self-incurred tutelage is the freedom to use one's reason critically and openly. Furthermore, for Kant it must be understood that this freedom is recognized only in terms of the transformation of subjective rules into principles that apply universally. Kant states this point explicitly: "The public use of one's reason must always be free, and it alone can bring about enlightenment among men."[20]

For Kant, reason's engagement of itself defines the process of critique; this much is clear. Even though it pursues a specific end, enlightenment, it is a perpetual process. According to Kant, the eighteenth-century is not "an *enlightened age*"; it is, however, "an *age of enlightenment*." In other words, as far as Kant is concerned, eighteenth century thought still lacks the independent use of critical reason. But there are "clear indications that the field has now been opened wherein men freely deal with these things and that the obstacles to general [universal] enlightenment or the release from self-imposed tutelage are gradually being reduced."[21] In this respect, Kant's attempt to realize the ideals of this process—both speculative and practical, or both subjective and universal—culminates in three questions.

1. What can I know?
2. What ought I to do?
3. What may I hope?[22]

Given the universal application of science and the universality of the enlightenment process, Kant's questions become questions for everyone. They invite everyone to engage in the pursuit of wisdom and self-knowledge in terms of the clarity and certainty required of and by critical reason. These questions provide the framework within which enlightenment can come about. Stated in a more explicit manner, this framework is identified by Kant as the "culture of human reason."[23] As a culture whose main characteristic is the play or spontaneity of reason, it provides the grounds for hope. It holds out the promise of scientific knowledge, the promise of universal laws, and the promise of ethical norms as Ideals that can be pursued and realized. Thus, it is not only the promise of knowing nature, in the sense that Bacon and Galileo thought nature could be known. Nor is it only the promise of knowing the self in order to establish rules of conduct, in the sense that Hobbes thought possible. It is the promise of finding those values that will guide everyone in whatever endeavor they are involved, in any arena in which they might act.

To be sure, Kant's picture of enlightenment supports the testaments and legends of optimism associated with the Enlightenment. Enlightenment is a process that promises liberation through education. *"Popular enlightenment* is the public instruction of the people upon their duties and rights towards the state to which they belong."[24] Individual enlightenment cannot remain, for Kant, separate from the enlightenment of society as a whole. Only in terms of a culture of enlightenment does it make sense, in Kant's texts, to talk about individuals as "ends" and not as "means," and therefore to talk about society as "a kingdom of ends."[25] Society, or a "republican" state, to use Kant's phrase, "should treat the people in accordance with principles akin in spirit to the laws of freedom which a people of mature rational powers would prescribe for itself, even if the people is not literally asked for its consent."[26]

Kant recapitulates his optimistic vision of what can be achieved through enlightenment when he stresses that freedom in thinking will eventually affect "the principles of government, which finds it to its advantage to treat men, who are now more than machines, in accordance with their dignity."[27] As the critique of reason already indicates, for Kant human dignity is not to be identified with the mechanical and routine operations of thought. Instead, it is to be identified with the spontaneity of reason—the ability to create freedom and to will laws. Individuals are autonomous rational beings. So, individuals are free to stipulate their relations to nature and to one another, even when it is an issue of the perfectability of nature or society.

Nature has willed that man should produce entirely by his own initiative everything which goes beyond the mechanical

*ordering of his animal existence, and that he should not
partake of any other happiness or perfection than that which he
has procured for himself without instinct and by his own
reason.*[28]

Rousseau: The Deception of Enlightenment

It is clear that Kant's questions—concerning what can be known, what
one ought to do, and what one can hope for—prefigure his attempts to
articulate the boundaries of reason and sensible experience. Further,
these questions set the stage for constructing a system of knowledge
that leads to enlightenment. Kant's elaborate system is eclipsed by a set
of specific values. The questions are formulated in such a way as to
presuppose their answers. That is to say, the answers that satisfy the
questions must be faithful, at least, to: (1) the possibility of objective
knowledge; (2) the demand that one perform one's duties in accordance with universal laws; and (3) the belief that the knowledge of
nature and self is knowledge of the "perfectablity" that inheres in
nature. These values constitute specific ideals that can be realized only
in terms of the faith Kant expresses in the authority and power of reason. In the spirit of enlightenment, reason legislates both the laws of
nature and the laws (and duties) that guarantee harmonious social
order.

It is exactly this appeal to the authority of reason, as if reason were
the universal tribunal for determining the source of enlightenment,
that Jean-Jacques Rousseau challenges when he responds to the question posed by the Academy of Dijon in 1750. The question is: "Has the
restoration of the sciences and arts tended to purify or corrupt
morals?"[29] As the discussion of education in *Emile* indicates, this is a
question central to the organization of Rousseau's texts. But beyond
orienting Rousseau's texts, it is a question that bears a legacy of the Renaissance and, as such, is central to any text concerned with the notion
of *enlightenment* in the eighteenth century. Kant would answer this
question in an affirmative fashion. For him, the arts and sciences have
not only purified morals but have shown how moral conduct can with-
stand reason's critique.

Rousseau, on the other hand, decries the "ornamentation" of reason, the "deceptive uniformity" of mental and social customs, and reason's presumptuous authority regarding knowledge of nature and self.
Rousseau's treatment of scientific theories in *Emile* articulates his insistence that the significance of any kind of knowledge rests in its use.
Knowledge—or the means of creating it—possesses no inherent value.

In its constitutive functions, as found in Kant's critiques, reason legislates its value and use, in accord with certain principles. For Rousseau, reason can at best only regulate, that is provide practical guidelines for its application. It is in this sense that Rousseau understands reason—or the sciences and arts—as an instrument that becomes a powerful and dangerous "weapon" for protecting a certain set of ideals. He says, "nature wanted to keep you from being harmed by knowledge just as a mother wrests a dangerous weapon from her child's hands; . . . all the secrets she hides from you are so many evils from which she protects you. . . ."[30]

To understand Rousseau's response to the question as it is posed, one must recognize a certain rhetorical turn in the language of the discourse. There is a definite shift in emphasis placed on the value of reason and its products (the sciences and arts). The question posed by the Academy of Dijon presumes that reason possesses an inherent value. But as Rousseau's discourse illustrates, the issue is not about the inherent value of reason, nor about the inherent value of the sciences and arts. It is instead an issue of the historical deployment and use of the sciences and arts as they pertain to the development of morality. He asks the Academy, and his readers indirectly, to "reconsider, then, the importance of your products; and if the works of the most enlightened of your learned men and our best citizens provide us with so little that is useful, tell us what we must think of that crowd of obscure writers and idle men of letters who uselessly consume the substance of the State."[31]

As far as Rousseau is concerned, by posing a question about the contribution of the sciences and arts to the purification or corruption of morality, the Academy already recognizes the point of his discourse. In effect, the Academy acknowledges the authority the sciences and arts wield beyond the pursuit of "sublime knowledge." That is to say, the sciences and arts have been used, as Rousseau's historical references witness, as instruments for intellectual and social control. At one point in the discourse, Rousseau writes that "the daily ebb and flow of the ocean's waters have not been more steadily subject to the course of the star which gives us light during the night than has the fate of morals and integrity been subject to the advancement of the sciences and arts. Virtue has fled as their light dawned on our horizon, and the same phenomenon has been observed in all times and in all places."[32]

Just as Emile's tutor, Jean-Jacques, warns Emile, Rousseau cautions the members of the Academy against readily accepting the authority of scientific and literary texts. Education—moral or scientific—is not to be obtained through the study of theories and artificial constructs. As

Rousseau says in *Emile*, if there are any truths borne by nature, they can be known only through an individual's direct encounter with nature. In an analogous fashion, the "Discourse" asserts that if there are any truths and values borne by the pursuit of morality, they can be put into play only through a direct encounter with oneself in and through the individual practice of "virtue." Rousseau notes that the "good men" of Rome disappeared at the point where "learned men" began to appear. The advent of "study," the contemplation of virtue, results in the loss of the practice of virtue.[33] As the allusion to Socrates points out, a wise and "just man would continue to scorn our vain sciences; he would not help to enlarge that mass of books by which we are flooded from all sides; and, as he did before, he would leave behind to his disciples and our posterity no other moral precept than the example and memory of his virtue."[34]

Like Plato's Socrates, like Descartes, Rousseau insists that the best education is one that produces neither disciple nor master. "Those whom nature destined to be her disciples needed no teachers. Verulam [Bacon], Descartes, Newton, these preceptors of the human race had none themselves; indeed, what guides would have led them as far as their vast genius carried them?"[35] Indeed, here Rousseau announces his commitment to a particular legacy and legend of enlightenment.[36] Enlightenment is a process of liberation; a process of recognizing the forms in which knowledge appears and, thus, by which individuals come to deceive themselves. It is a liberation from the authority assumed by or presumed of any form of knowledge, regardless of the promises. According to Rousseau, "If a few men must be allowed to devote themselves to the study of the sciences and arts, it must be only those who feel the strength to walk alone" in the footsteps of Bacon, Descartes, and Newton and yet "go beyond them."[37] In other words, if there are any legacies or legends to be followed, they are ones that remain unwritten—they are to be written by *enlightened* individuals.

The salient features that characterize the process of enlightenment in Rousseau's texts appear deceptively similar to those announced in Kant's texts. Both Rousseau and Kant emphasize the autonomy of the individual, the independence of critical thought, and the desire to live as nature would command. However, Rousseau would object to Kant's insistence that reason is the mechanism for obtaining intellectual and moral freedom. Moreover, he would find suspicious the need to erect and support a uniform and universal set of ideals to guide the individual towards enlightenment. For Rousseau, then, one cannot be bound to any one system of thought (or critique) for the purpose of securing freedom. But there is more than a recognition of various critiques and approaches to enlightenemnt that traverses Rousseau's discourses. There is also a recognition of a multiplicity of possible enlightenments, that is a multiplicity of ideals that may be worth pursuing even if these ideals do not fit with the goals of Enlightenment.

Kant's critiques constitute a discourse on the paths one should and can follow in order to pursue enlightenment. Furthermore, Kant's critiques map out certain other paths that are excluded in his efforts to determine the perimeters of enlightenment. By contrast, Rousseau's discourse constitutes a critique of Enlightenment. According to Rousseau, as articulated and pursued in certain scientific, ethical, and philosophical texts of the eighteenth century, Enlightenment requires accepting and adhering to a certain set of paths mapped out in advance, a priori. Where Kant indicates the boundaries beyond which reason fails because it cannot comprehend all possible paths, Rousseau forces a remapping of the plan for enlightenment by attempting to trace those paths that might be otherwise left unblazed and unrecognized. Both Kant and Rousseau recognize possible passageways that drift in directions other than the one(s) marked out by the Enlightenment. In this regard, both Kant and Rousseau recognize the complex play and incessant generation of images and precepts underlying enlightenment. Moreover, both Kant and Rousseau acknowledge that the activity and artifacts of enlightenment involve embracing an intricate network of fictions, that is to say a labyrinth. It is, however, a labyrinth that does not indicate clearly which—if any—path or paths are to be preferred or for what reasons. Nor does it indicate how or if the play of images and artifacts, that is the labyrinth itself, can be comprehended.

Hume: The Enigma of Enlightenments

Enlightenment is, most definitely, an issue of a labyrinth in the texts of David Hume. In a fashion more explicit and elaborate than either Kant or Rousseau, Hume directs attention to and explores the enigmatic character of knowledge and understanding. He closes The Natural History of Religion [1757] with the following passage:

> The whole is a riddle, an aenigma, an inexplicable mystery.
> Doubt, uncertainty, suspence of judgment appear the only
> result of our most accurate scrutiny, concerning this subject.
> But such is the frailty of human reason, and such the
> irresistible contagion of opinion, that even this deliberate
> doubt could scarcely be upheld; did we not enlarge our view,

and opposing one species of superstition to another, set them a quarrelling; while we ourselves, during their fury and contention, happily make our escape into the calm, though obscure, regions of philosophy.[38]

For Hume, the "frailty of human reason" is characterized by the "natural propensity" of reason to systematize; that is the desire to impose an order on the universe, to present a system—to create a "whole"—out of a multiplicity of natural phenomena. Reason is, according to Hume, "nothing but a wonderful and unintelligible instinct in our souls, which carries us along a certain train of ideas, and endows them with particular qualities, according to their particular situations and relations."[39] The frailty of reason is, then, the frailty of a natural instinct: it is unable to maintain the order it imposes on nature and itself. It buckles under the weight of its edifices because the internal alignment of these structures is always shifting. When reason attempts to present a universal system "the ideas, which are the parts of its discourse, arrange themselves in a certain form or order; which is not preserved entire for a moment, but immediately gives place to another arrangement."[40] The discourse of reason—what is the play and configuration of perceptions in Hume's texts—inaugurates a self-imposed displacement of structures, eliciting the exception to its rules and principles.

According to Hume in *A Treatise of Human Nature* [1738-1741], nature is perceived at once as a unity and a multiplicity. The first line of the *Treatise* sets up the general organizing principle of Hume's discourse on knowledge and, thus, the labyrinth of fictions constituting his texts. Hume begins: "All the perceptions of the human mind resolve themselves into two distinct kinds, which I shall call Impressions and Ideas."[41] As with other general principles articulated in Hume's text, this principle is presented for the purpose of signaling the exceptions to its rule. The distinction between impressions and ideas is defined in terms of the differences in the "force," "liveliness," and "violence" by which each kind of perception "strike[s] upon the mind." According to the initial statement of this principle, impressions are more forceful than the "faint images" constituting ideas. Ideas are, then, "copies" of impressions whose force remains intact after the impression has vanished.

Hume's "first principle" in the "science of human nature" engenders another general division in the classification of perceptions that covers both impressions and ideas. "This division is into Simple and Complex."[42] While simple perceptions permit no division as such, complex perceptions are distinguishable. Particular colors, smells, and boundaries, for example, that appear united in a forest can be separated from each other. That Hume is content establishing "one general proposition" regarding the apparent correspondence between impressions and ideas betrays a more complex set of circumstances. The "rule" that "ideas and impressions appear always to correspond to each other," Hume readily acknowledges, is "not universally true." Ideas and impressions are not "exact copies of each other," even though *simple ideas in the first appearance are deriv'd from simple impressions, which are correspondent to them, and which they exactly represent.*[43]

A second glance reveals a more complex division of perceptions. Moreover, it reveals the two faculties governing this division. Impressions are of two sorts: impressions of sensation and impressions of reflection. According to Hume, the first, impressions of pain or pleasure, "arises in the soul originally, from unknown causes"; the second, passions, desires, or emotions like hope and fear, are "derived in great measure from our ideas."[44] In order for an impression to re-present itself—that is to say repeat itself and, thus, present an image of itself—, it does so in one of two ways: memory or imagination. Memory permits an impression to present itself again in a form that "retains a considerable degree of its first vivacity." In this respect, the impression is neither wholly an impression nor wholly an idea; it is, as Hume indicates, somewhere in between. Imagination, on the other hand, enlivens an impression so that it may be repeated when its initial vivacity has been lost and when it has become "a perfect idea."[45] For the purposes of the immediate discussion, Hume's treatment of memory will provide the focus.

Where the order and position of impressions and ideas are integral to the understanding of an object or phenomenon, memory is already at work. As Hume notes, "The chief exercise of the memory is not to preserve the simple ideas, but their order and position."[46] Memory is one faculty involved in the association of ideas, or the set of relations between ideas Hume classifies under the headings of resemblance, contiguity in time or space, and cause and effect.[47]

In terms of these relations, the association(s) of ideas constitutes the discourse and the economy of reason. To be certain, this discourse and economy reflects reason's desire to build a system.[48] On the basis of the "impressions and ideas of the memory we form a kind of system, comprehending whatever we remember to have been present, either to our internal perception or senses. . . . " Once a system is in place, "every particular of that system, joined to the present impressions, we are please'd to call a *reality*."[49] However, with the creation of the system, and as such reality as a whole, there is no end to the fabrication. One system, or in Hume's language, one impression or one fiction, is always associated or connected, by custom — "or if you will, by the relation of cause or effect" — , with another system/fiction. Any "new system," determined by the convergence of an old system with yet a different one, is dignified "with the title of *realities*."[50]

Even within the limited use and range of memory (as opposed to "the liberty of imagination"), reality — nature, self, and society — is produced, fabricated if you will, through the discourse of reason, that is to say the interplay of ideas and impressions. According to Hume, reality is a fiction, but a fiction that is not abstracted from a simple idea. On the contrary, reality is a complex nexus of fictions. Like any other fiction bound to a particular system of ideas, reality traverses the labyrinth within which it arises. But the principles on which a system rests, and this certainly applies to Hume's systems as well as the systems of science, religion, politics and morality, are articulated only once the system is in place and at work.

As far as Hume is concerned, principles are created, or chosen as some might say, in light of the purposes for which they are to be used. That is to say, "It is very natural . . . for men to embrace those principles, by which they can best defend their doctrines. . . . "[51] In religious matters, for example, "whichever system best suits the purpose" of the clergy, "in giving them an ascendant over mankind, they are sure to make it their favourite principle, and established tenet."[52] Or in political matters, as Hume contends,

> As no party, in the present age, can well support itself without
> a philosophical or speculative system of principles annexed to
> its political or practical one, we accordingly find that each of
> the factions into which this nation is divided has reared up a
> fabric of the former kind in order to protect and cover that
> scheme of actions which it pursues.[53]

And finally, when discussing moral systems in *An Enquiry Concerning the Principles of Morals* [1751], Hume focuses on the arguments and disputes over moral distinctions. Hume remarks in a rather cynical tone that "as reasoning is not the source, whence either disputant derives his tenets; it is in vain to expect, that any logic, which speaks not to the affections, will ever engage him to embrace sounder principles."[54]

Reason, its principles and its discourse, functions within a self-generating abyss of representations; an abyss generated through instinct or custom. Religious beliefs, political party lines, and moral systems plot at least three of the more "vulgar" domains or systems situated within the abyss. The systems of philosophy and science are part and parcel of the abyss as well. Given the novel character of reality, it is not surprising to find that the principles underlying each of these domains are derived in analogous ways. They are produced post hoc: they are supposed to give the appearance of a methodological apparatus that ensures consistent application of a system. The myths of clarity and certainty and the principle of reason that inform the "scientific method" of the seventeenth century are no exceptions. They too are produced to gird a metaphysical system, religious dogma, or scientific enlightenment already in place, already engaged.

Contrary to the beliefs that orient certain philosophical systems and scientific texts of the seventeenth and eighteenth centuries, by regarding reason as a "natural instinct" Hume is in the position to claim that the products of reason—"reasonings," "judgments," and "inferences"—are "the effects of custom."[55] Custom or instinct can do very little on its own except when it enlivens an idea or impression by way of imagination or memory. Furthermore, he is in the position to advance two different principles that can guide any inquiry as it threads its way through the labyrinth of reason's faculties and, thus, attempts to comprehend the sublime limits of this labyrinth. Hume writes:

> That there is nothing in any object, consider'd in itself, which can afford us a reason for drawing a conclusion beyond it; and, That even after the observation of the frequent or constant conjunction of objects, we have no reason to draw any inference concerning any object beyond those of which we have had experience.[56]

To be certain, it is on the basis of natural inclinations—mental habits and appetitive sensibilites—that inferences and judgments are generated. So, where Kant believes that the faculty of judgment follows a schema prescribed by the categories and is, thus, universal in its functions, Hume resists the proposition that all judgments and inferences are produced in identical habitual forms. As Hume notes, "All depends on our manner of viewing the objects."[57] In other words, one must take into account the context, and the different ways in which one is engaged in that context, in order to understand the differences in judgment and perception that might be expressed.

Guided by these two principles, knowledge for Hume can be understood only in terms of probabilities. The traditional distinction between knowledge with certainty and probable knowledge dysfunctions when considered in light of the association of ideas and impressions. Hume's principles show that even though there is no objective feature that warrants drawing a specific conclusion about an object, reason nevertheless pursues the establishment of such a conclusion. Moreover, when reason moves towards a conclusion or a judgment concerning the connection or relation between objects, it does so solely on the basis of the frequency and vivacity with which these objects appear related. When objects appear connected in a definite fashion over a certain period of time, reason—by way of imagination or memory—supplies the connection. The frequent appearance of objects in a specific order and position does not guarantee they will appear in that form in the future. Reason, though, is disposed to presume regularity—that is to say, reason presumes the resemblance of the past and the future.[58]

The general principles used in Hume's analysis of knowledge, belief, habit, and custom function as signs indicating both familiar and unfamiliar paths towards enlightenment. Of course, as with Rousseau's discourse, some of these passages remain uncharted. Hume's general principles serve as means for simplifying the intricate network of phenomena encountered by the faculties of reason. As such, they differ in their legislative capacities from Kant's rules (concepts) of pure understanding. Unlike Kant, Hume is not involved in establishing the universal, a priori conditions that make knowledge or experience possible. He is, however, engaged in a description of the techniques used in establishing what can be known about the present and the future with some regularity. Herein lies Hume's engagement of the labyrinth and the source of his enlightenment(s): describing the techniques (fictions) used to erect systems of belief involves or comprises the performance of those very techniques.

The enigmatic character of reality—the whole, as Hume says, or the self, or nature—forces Hume to articulate certain principles to guide his inquiry. As Hume readily acknowledges, the labyrinth of fictions cannot be comprehended—as a whole, as a system. It is not a system of reflexion that is able to capture itself or comprehend itself in its own medium. Acknowledging this point is in itself a moment of enlightenment. Hume understands that any comprehension of reality, nature, or self is at best provisional and fictional: he understands that any such comprehension is ad hoc. Our understanding of the whole always changes in light of our activities that unfold and thus define its boundaries. Adherence to a theory or a single set of principles that supposedly delimit the labyrinth and make it comprehensible is, in effect, an involvement in a incessant generation of fictions. Fabrication, then, provides the only support for understanding, explanation, and relief from doubt.

If reality, nature, or the self is enigmatic, then understanding is, always and already, incomplete. Thus, fabrication supplements the activities of knowing by providing the necessary techniques that will complete the process of understanding. The realization that the techniques (or critiques) of reason are themselves fictions is a second moment of enlightenment in Hume's texts. These techniques are created in and through their use, and therefore differ from those rules of reason described by Kant as natural laws. In this respect, Hume recognizes the duplicity of enlightenment. In its desire to comprehend the whole of reality, which is taken to be a riddle, reason substantiates the enigmatic character of the whole through the construction of its systems.

The techniques of reason are interlaced with the enigmas they attempt to unravel. The multiplicity of techniques deployed by reason in its efforts to comprehend an object as such provide the clearest indications of reality's enigmatic character according to Hume. Furthermore, the multiplicity of techniques parallels the multiplicity of perspectives brought under the principles of philosophical or scientific systems. A system, as Hume's *Treatise* demonstrates, acts as a prism through which a multiplicity of realities is revealed. Despite the order and regularity provided by any system, Hume announces the impossibility of reason coming to rest with any particular point of view in advance of its engagement in a particular context. The frailty of reason is understood in terms of its vascillation between "opposite views."[59] Such is a third moment of enlightenment in Hume's texts. Certainty is displaced by "opinion"—the confidence of self-legitimating discourse, or narrative–fictions.

However, for Hume this third sort of enlightenment may provide no enlightenment at all: vascillation arises from and entails confusion, bewilderment, and disquietude. It is possible that no one passage would be recognized as different from any other passage in the labyrinth. There is, then, a fourth moment of enlightenment. When one chooses a particular path, it is in accordance with a given set of concerns or in order to achieve a particular purpose. Not only do individuals recognize at some point in their deliberations that they act within a labyrinth, but they are struck by the need to chart the path most appropriate for the moment, in spite of what any system dictates and what they are willing to confess given their adherence to a system. Trailblazing becomes a form of enlightenment.

Impressions of the Enlightenments

Within the complex economy of exchanges constituting what is called the history of philosophical and scientific discourse, the texts of Kant, Rousseau, and Hume are relays or terminals through which certain enlightenment attitudes and themes become evident. Kant's texts announce a spirit of optimism associated with the enlightenment of reason; Rousseau's express a sense of distrust in the rule of reason imposed on individual thougth; and Hume's present a sense of ambiguity, where reason fluctuates between the certainty of optimism and the doubt of skepticism and pessimism. But what matters, at least for the present discussion, is not the determination of *the enlightenment spirit* to be found in any of these three systems. Seen as relays, the texts of Kant, Rousseau, and Hume are not points of termination. Moreover, there is no final, unique point to be gleaned in their articulation. Instead, they are sites for arrival and departure, issuing certain textual themes and lines of thought. They provide another series of pegs, listening posts, or techniques by which the legacies and legends of enlightenment are transmitted. As such, the significance of the Kantian, Rousseauian, and Humean texts lies in the ways in which the impressions/ideas they issue are appropriated in other critiques (or techniques) of reason, especially regarding the progress of science and technology.

The texts of Karl Marx and Friedrich Nietzsche are discussed in this context not for the sake of providing some historical continuity, nor for the sake of documenting the ongoing influence of the texts of Kant, Rousseau, and Hume. Instead, they illustrate in explicit ways the interlacing of attitudes regarding science and technology that take shape during the eighteenth and nineteenth centuries and that continue to prejudice us today. Although the concerns articulated by Marx and Nietzsche, as well as the issues treated by Kant, Rousseau, and Hume, cannot be reduced to the many questions and themes dominating current discourses on science and technology, their texts continue to narrate contemporary discourse in several very subtle ways.

Taken as an ensemble, these texts make up a series of relays, passages, or techniques for interpreting and thus creating certain legacies and legends regarding the cultural role of science and technology. But what is being relayed? The techniques by which the world is created, viewed, and controlled. That is to say, all that is relayed are the relays themselves.

Marx: The Ambiguity of Critique

Karl Marx's texts are marked by an ambiguous appropriation of certain enlightenment themes. At a very general level, Marx embraces the Kantian/Hegelian belief that reason furnishes the techniques required in the critique of consciousness. "Consciousness is," according to Marx, "from the very beginning a social product, and remains so as long as men exist at all."[60] Like the production of ideas and conceptions, the production of consciousness is "directly interwoven with the material activity and the material intercourse of men—the language of real life."[61] More explicitly, language sets the boundaries of consciousness through the production of ideas, theories, and systems of order. For Marx this critique focuses on the historical, material, social, and linguistic conditions that inform an individual's awareness of self and the understanding of those conditions as they pertain to individual activities.

The task of critique is to expose and eliminate the fictions that control the speculation about these conditions. As Marx writes, "The immediate *task of philosophy* [reason or critique], which is at the service of history, once the *holy form* of human self-estrangement has been unmasked, is to unmask self-estrangement in its *unholy forms*."[62] Unmasking the ways in which individuals are dominated by certain ideas or images of themselves, and their relations to others and nature, is tantamount to a kind of enlightenment for Marx. Like Kant and Hegel, Marx recognizes that criticism or enlightenment cannot be conceived of as an end in itself. Criticism is a technique of practice (*praxis*); a means for realizing certain ideals absent from the existing order.[63]

Marx's appropriation of reason's critical function and task is given an ambiguous articulation. Critique takes place on several different although congruent levels at once. Moreover, as it is used in the pursuit of particular ends, critique simultaneously assumes two attitudes: pessimism and optimisim.[64] Marx's critique of speculative reason is dominated by his concerns with reason's ability to affect the conditions of human existence, especially in light of the historical use of reason. But the pessimism of Marx's critique is used for another end: the construction of a new system of critique; and here Marx's optimism becomes evident. Whatever tone Marx assumes in his texts, it is clear that critique is never indifferent. Marx insists on the practical deployment of critique. It must make a "radical" break with tradition, at one point, and yet to do so it must employ the rational techniques handed down through the tradition.

Marx criticizes Kant and Hegel for withdrawing reason from its engagement in the practical. In the writings of these "critical theologians," a group in which Marx includes Kant, Hegel, Feuerbach, and Bauer, the critical function of reason never moves beyond the speculative dogmas of theology.[65] In the tradition of pure reason, reason accepts its own presuppositions as authoritative, as legitimating its abstract reign. Thus, criticism ends in a *"speculative cycle*, and thereby [in] its own *life's career. Its* further movement is *pure*, lofty *circling within itself*, above all interest of a *mass nature"* and devoid of any interest in its application to social relations.[66] In light of speculative reason's circularity, Marx launches an attack on or a critique of classical political economy. The circularity of speculative reason obfuscates what Marx takes to be actual economic conditions. The abstract categories of labor, capital, wages, and profits are presented in the theories of David Ricardo and Adam Smith, for example, as if they have assumed a life of their own, void of any empirical reference. Ricardo and Smith presume the reality of private property, without explaining its historical and material origins. These categories, according to Marx, are used by classical political economists[67] and the bourgeoisie to justify what is a set of unjustifiable economic conditions and relations.[68]

Yet Marx does not abandon the faculty of reason. Apposed to the pessimism Marx expresses regarding reason's role in the science of political economy, he finds embedded within the classical critiques reason's radical moment of promise.[69] He elaborates the classical enlightenment conception of reason's power to protect old systems as well as construct new ones. (Indeed, in a spirit reminiscent of Kant's three critiques, or Hegel's *Encyclopedia*, the three volumes of *Capital* constitute a new comprehensive and critical system for understanding the relation of self, nature, and society.)

Like Rousseau, Marx believes that to the extent that reason, theory, and criticism, of the speculative sort, produce the techniques of control and domination of individuals, they are equally a "weapon of criticism," a "material force" of the system (of a text or any analogous labyrinth) that can be used for the liberation of individuals. According to Marx, "Theory is capable of gripping the masses as soon as it demonstrates ad hominem, and it demonstrates ad hominem as soon as it becomes radical. To be radical is to grasp the root of the matter. But for man the root is man himself."[70]

The radical turn in Marx's theoretical apparatus results in a vision of *"Man"* as "the immediate object of natural science" and *"nature"* as "the immediate object of the *science of man*."[71] The return to "man" or human beings (as a species) is inextricably tied to the spirit of the

Kantian, Rousseauian, and Humean enlightenments. Individuals are involved actively in the production of themselves. They create the means, the techniques, and the strategies for imposing order on nature, self, and society. And this production is accomplished through the production of the "conceptions, ideas, etc.," of which individuals are the sole authors.[72]

The production of theories and systems, in other words the "*natural science of man,*" is itself a social activity. It describes the conditions under which individuals determine and are determined by the products of their interactions with one another and the artifacts of their actions. Theory, then, must have an empirical grounding; it must take the form of science. But beyond the reference to positive material conditions and social relations, science, scientific inquiry in Marx's terms, is conditioned by language. Language's significance, for the constitution and use of science, rests on its being the medium of scientific consciousness.[73] Language is "practical, real consciousness,"[74] and, as such, is the only means for articulating and deciphering "the secret of our own social products."[75]

Once Marx makes this announcement, it is easy for him to state in a very Hegelian manner, for example, that "respective religions are no more than *different stages in the development of the human mind.*"[76] The science of man, as far as Marx is concerned, can free individuals from abstract institutional and linguistic categories. That is, "the relation of Jew and Christian is no longer religious but is only a critical, *scientific* and human relation."[77] Just as science can free individuals from the self-alienation of speculative religion, it can identify the unity of human conditions and relations that constitutes the individual as such.[78] And if any contradictions become apparent in the critique of these conditions, they do not arise from any inherent contextual oppositions. Therefore, according to Marx, they can be resolved—explained, understood, interpreted—"by science itself."[79]

Throughout each of his critiques, Marx attempts to recover the critical "rational kernel" concealed by the "mystical shell" of speculative philosophy. In order for a critique of reason's deployment to comprise a science, it must outline the possibilities of its application. Science, for Marx, is always radical and emancipatory. Marx's vision of enlightenment is not restricted to the liberation of individuals. Instead, science enlightens and liberates only when the breadth of its application encompasses all classes of individuals and the material, historical, and social conditions of their existence. Despite his protests to the contrary, Marx echoes Kant's desire to establish society as a "kingdom of ends." Philosophical, political, and social revolution is directed toward the liberation of society as a whole, that is the elimination of class struggle; liberation comes about only when the material conditions oppressing the masses are transformed.

If science identifies the conditions of oppression, as well as the possiblities for change, then technology defines the medium for displacing those modes of production and those relations that maintain the conditions of alienation, exploitation, and servitude. But the ambiguity of Marx's critiques remains: to the extent that technology creates the conditions of liberation within a particular system, it intensifies the conditions of oppression. The difference between these two possibilities depends on the *use* of technology and science, that is to say the ends toward which technology and science are geared.

Technology in this context, according to Marx, is "the application of . . . science to production." Marx recognizes that technology is not merely applied science. There is a relation of reciprocity that defines the respective domains of science and technology. "Natural science," as Marx says in the notebooks for *Capital*, "is itself in turn related to the development of material production."[80] The reference to "modes of production" or "material production" is linked to Marx's understanding of modern industry. Where science and technology are determined reciprocally, modern industry also stands in a reciprocal relation to science and technology—a relation in which industry provides the material basis for the development of science and technology. Given this dynamic and intricate view of science, technology, and industry, Marx claims that "the technical basis of that [modern] industry is therefore revolutionary." Because modern industry never accepts "the existing form of a process as final," it continually forces "changes not only in the technical basis of production, but also in the functions of the labourer, and in the social combinations of the labour-process."[81]

In his treatment of science and technology, Marx transcends the metaphysical concerns of methodology. Instead of sharpening his critique by adhering to one of the enlightenments of the eighteenth century, Marx links the development of science and technology—as well as philosophy and political economy—to a practical domain: modern industry. Though Marx is as concerned as Kant, Rousseau, and Hume with the question of "order," for him the ordering of nature, self, and society is affected through the material conditions of production (division of labor, capital formation, and the use of machinery). Marx remains concerned with the issues of order and progress; but he stresses a different kind of order, a different mode of progress. He focuses on the internal alignment of labor, capital, and machinery as the organizing factors of modern industry, and not on the establishment of a particular specualtive system. According to Marx, if science and technology are to enlighten and liberate, in other words if they are to make a difference in the workings of modern society, they must offer specific suggestions for transforming the work place based on the existing means of control and alientation.

Yet Marx's recognition, that the critical deployment of reason through science and technology bear the means of liberation, arises from the recognition that science and technology also provide the means for domination and control. Any system of production, any form of organization, or any ruling principle betrays the means of its own collapse and negation and thus its own displacement. The use of science and technology, for already identified purposes within a specific context, determines the difference between liberatation and domination. *Artifice* determines which paths evoke a sense of liberation or a sense of continued exploitation and alienation. Science and technology, or the techniques of reason as it were, can be used to reduce or exacerbate human oppression.

> The implements of labour, in the form of machinery,
> necessitate the substitution of natural forces for human force,
> and the conscious application of science, instead of rule of
> thumb. In Manufacture, the organization of the social labour-
> process is purely subjective . . . in its machinery system,
> Modern Industry has a productive oragnism that is purely
> objective, in which the labourer becomes a mere appendage to
> an already existing material condition of production.[82]

Here Marx notes a shift from the "subjective" organization of production, where production is controlled by the worker, to an "objective" organization of production, where production is managed by machinery. The shift is duplicitous; it signifies both a positive (or optimistic) and a negative (or pessimistic) transformation. On the one hand, reason dominates through its instruments, i.e., the techniques of organization and management. The use of reason promises unity of thought and action: consensus. Order and regularity are guaranteed. By contrast, in the shift from the subjective to the objective, the individual "becomes a mere appendage" of the technological and mechanical apparatus. Creativity is lost through order and regularity; differences in perspective—the interplay of thought and action—are masked by consensus.

Just as machines create the potential for freeing individuals from arduous tasks, from the heavy and unpleasant burdens of manual work, they also create the conditions that bind the life of the worker to the life of the machines. The machine appears to take on a life of its own, consuming every aspect of an individual's identity. Here Marx echoes the dire descriptions cited in Charles Dickens's *Hard Times*, "where the piston of the steam-engine worked monotonously up and down like the head of an elephant in a state of melancholy madness."[83] Again, Marx contrasts the positive expectations of efficiency and productivity that result from the use of machines—the shortening of produciton time—with the negative impact on individuals who work with machines—mechanization of industry extends the work day to its fullness. In *Capital* Marx writes:

> If machinery be the most powerful for increasing the productiveness of labour—i.e., for shortening the working-time required in the production of a commodity, it becomes *in the hands of capital* the most powerful means, in those industries invaded by it, for lengthening the working-day beyond all bounds set by human nature.[84]

Nietzsche: The Engagement of Ambiguity

Embedded within Marx's critique of reason, one can trace an appreciation for the role language plays in constituting the context or labyrinth of individual activity. Marx's formulation of the natural science of man presupposes a certain view of language. Language is a product of social relations. As such, language is bound, from the outset, to the practical. Indeed, the few scant references to language, found within Marx's texts, indicate that for Marx language is the medium for the production of the practical as such. It is the medium for the creation of all social relations and modes of production.

The creation of ideas and concepts is an integral element in the fabrication of social relations and modes of production, and the creation of consciousness. Consciousness is presupposed in the construction and articulation of social relations. The interaction between theory— radical critique or science—and praxis—social activity—is, then, the simultaneous creation of consciousness and reality. And the representation of this interaction, in and through language, is only a manifestation of actual circumstances. Neither theory nor social relations, nor the medium of their presentation—language—, form independent realms of their own.[85] Their historical interaction orders the systematic organization of a labyrinth as well as maintains its enigmatic character.

Marx's texts are caught up in a labyrinth of fictions that they themselves attempt to systematize. In a subtle way, Marx seems to appreciate, and acknowledges, this inescapable quandry.[86] But because of the desire to articulate a new philosophical system, or a new "science" (a la Kant and Hegel), the ambiguity that marks the use of language, through which Marx's enlightenments are "actualized," is deferred. Again, because of the desire to systematize, in accordance with particular legacies, Marx cannot afford to expose his enlightenments to ambiguity. Marx cannot acknowledge explicitly the extent to which ambiguity permeates the foundations of his critique. Such an acknowledgment would rob Marx's critique of its value as the replacement for other critiques. In spite of, and perhaps because of, Marx's intention to create a system that would replace all others, his system becomes one more passage in the labyrinth of enlightenments. Marx's enlightenments engender displacement, not replacement.

Friedrich Nietzsche's texts celebrate the ambiguity of recurrent displacement that haunts any critique. The venue of this ambiguity is interpretation—the labyrinth of language. According to Nietzsche, the classical notions of a "timeless knowing subject," "pure reason," "absolute spirituality," and "knowledge in itself" are pure fictions on which rests the security of understanding or, more generally, the security of any system.[87] These fictions provide the sum of observations on which any system would be grounded. As such, these observations are interpretations of a world that, for Nietzsche, is in a constant state of becoming. *Interpretations* are always signs of *perspectives*, engendering multiplicity and difference. In *On the Genealogy of Morals* Nietzsche writes that:

> There is *only* a perspective seeeing, *only* a perspective
> "knowing"; and the *more* affects we allow to speak about one
> thing, the *more* eyes, different eyes, we can use to observe one
> thing, the more complete will our "concept" of this thing, our
> "objectivity," be.[88]

In short, the "world," "life," "being," the products of any pursuit of knowledge remain falsehoods that can never betray or disclose *the* truth or any truths.

Interpretation *fixes* the world; it makes the world what it is; it determines the "character of the 'appearance'!"[89] Such a determination forges the domination of a particular perspective or interpretation.[90] All interpretation is a translation of the world—as it unfolds—into a language that makes the world common and accessible. Without the constraint of language, then, there would be no concept of the world, life, meaning, or knowledge. "Of all the interpretations of the world attempted hitherto, the mechanistic one seems today to stand victorious in the foreground."[91] According to Nietzsche, mechanistic "procedures" or "scientific method" ensure that science can "achieve progress and success."[92] Here Nietzsche expresses an enlightenment theme found in Hume's *Treatise*. The theory of mechanics provides certain philosophical systems with principles that support the explanation of natural phenomena. But, as Hume notes, these explanations are derived from analogy—the association of natural events and the impressions of regularity they register by resemblance. Resemblance connects impressions and ideas with an interpretive framework within which the translation is carried out. In Nietzsche's words, " 'Regularity' in succession is only a metaphorical expression, *as if* a rule were being followed here; not a fact."[93]

Concepts orient thinking, but they do so as metaphors and not as categorical imperatives. Concepts/metaphors constitute the means that orient thought and activity within a particular perspective. The multiplicity of perspectives demonstrates the absence of correspondence between conceptual framework and the structure of reality. Instead, for Nietzsche, the only structure to be encountered is metaphorical. The most intricate metaphorical structure explored by Nietzsche is delineated in his genealogical/etymological account of moral concepts/principles and the ascetic ideal.[94]

Nietzsche abandons the legacy that traverses the texts from Bacon and Galileo up through Kant and Marx, dominated by the nostalgia for an "objective world" that can be represented or for the disclosure of secrets hidden in the appearances of nature. There is no deciphering of the text or the book of nature; there is no unique understanding, because there is no-*thing* to decipher nor comprehend. Moreover, "*the* way," through the text or out of the labyrinth of fictions, "does not exist."[95] There are only interpretations—translations of metaphors—expressing specific perspectives and values, creating the paths that constitute the labyrinth. For Nietzsche the only secret or enigma is coming to terms with the realization that whatever is called "truth" or "knowledge" is a fiction, one that has been interpreted as something else. As Nietzsche writes, "truths are illusions about which one has forgotten that this is what they are."[96]

To the extent that the words *understanding, knowledge,* or *reason* have meaning, the world is comprehensible or "knowable." "*Rational thought is interpretation according to a scheme that we cannot throw off.*"[97] But this moment of enlightenment is marked by a certain duplicity. The scheme is itself an interpretation, a product of the interpretation of interpretations. According to Nietzsche, any scheme of order, any organizing mechanism or method, is inextricably bound to the play of interpretations, the labyrinth of perspectives. The world, then, as constituted through the labyrinth of perspectives, does not countenance any unique vision or order because of the possibility that it *"may include infinite interpretations."*[98]

The critiques of reason, especially those of Kant, Hegel, and Marx, were intended to bring order to an already ordered world or nature. This is, as Nietzsche points out, the regulating principle that guides the solutions of the "world riddle" proposed by these enlightenment texts.[99] Except for Rousseau and Hume, these texts refuse to accept what is strange to their principles of order and, likewise, to acknowledge that whatever order is found is imported. In other words, these texts import what they purport to find; they create order through their own narratives—and this is what is called "science."[100]

All interpretation presupposes the ambiguity or enigmatic quality of life in Nietzsche's texts. Life is replete with contradictions, deceptions, and change. Science, as it is practiced from the perspective of its discourse, cannot function under these circumstances. Where one mode of interpreting attempts to narrow the scope of ambiguity, to reduce the depth of the abyss in which it finds itself, it does so quite arbitrarily. As one mode of interpreting, science generates "an artificial arrangement for the purpose of intelligibility— . . . by selecting one element from the process [of life] and eliminating all the rest."[101] Science is at odds with the world as "becoming," "as continuing and eternal"; to accomplish its tasks, science requires a world that is "finished and historical."[102] That is to say, science requires a truth that is fixed and eternal.

If the ideal of truth pursued by science transcends the limits of its own interpretive framework and methods, then it is pursued on the basis of faith. Indeed, the enlightenment of science is grouned in faith: the faith it has in itself, and the faith it instills in others about itself. It is, for Nietzsche, "still a *metaphysical faith*," a "faith in a *metaphysical* value, the absolute value of *truth*."[103] According to Nietzsche, this *illusion*—and nothing more substantial—constitutes the foundation on which science envisions the explicability of nature.[104] Thus, "this sublime metaphysical illusion accompanies science as an instinct and leads science again and again to its limits at which it must turn into art— *which is really the aim of this mechanism*."[105] Brought to its limits, science must transfigure them, recast them in order to accommodate the demands of its faith. It is at this point that science betrays its artistic foundations. Science is indistinguishable from art or *techné*: science and art are necessary correlatives.[106]

Does the realization that there is no exit from the labyrinth of fictions elicit the pessimistic legacy of the enlightenments? Does it evoke a form of nihilism? Nietzsche identifies nihilism as a form of *"weakness,"* just as he identifies Christianity, for example, with *"sickness."* The weakness of nihilism is expressed fully in the "unshakable faith that thought . . . can penetrate the deepest abysses of being, and that thought is capable not only of knowing being but of *correcting* it."[107] The nihilist lacks the strength to live within the labyrinth of interpretations. The world must be comprehensible, unified through one perspective to the nihilist. So, while nihilism forces the weak to adopt certain perspectives of which they are not the authors, it forces the strong to interpret, to create, author, and celebrate the ambiguities of life.[108] The strong affirm and enjoy the metaphorical labyrinth.

Is the consequence of Nietzsche's enlightenment to prefer art, to give it a position of primacy or authority in the order of human activity? Or, is it an attempt to grasp life as an aesthetic phenomenon, as a product of a specific genuis, where science and art are techniques by which life is affirmed—indeed, comprehended, justified, and found pleasant? Once the enlightenment of scientific inquiry is cast in terms of its linguistic pretext—the interpretive framework that endows it with meaning—its authority is suspect and thus suspended. The suspension of the legacies and legends of enlightenment is celebrated, it is an affirmation of the possibilities presented by the enigma—of enlightenments, of the labyrinth, of life. For Nietzsche, the task is to understand the enigma in terms of how science is grasped through the declarations of art, and how art is grasped through the declarations of life—interpretations.[109] The authority of scientific and technological discourse, then, is predicated on these linguistic performances.

Notes

1. Compare Jacques Derrida, "Déclarations d'Independance," in *Otobiographies: L'enseignement de Nietzsche et la politique du nom proper* (Paris: Éditions Galilée, 1984), 13-32.

2. Immanuel Kant, *Critique of Pure Reason* [1781 1st ed.; 1787, 2d ed.] trans. by Norman Kemp Smith (New York: St. Martin's, 1929), A xia.

3. Ibid., A xi-xii.

4. Ibid., A xvii.

5. Ibid., B xiii.

6. Ibid., A 51/B 75.

7. Ibid., A 80/B 106

8. Cf. Ibid., A vii.

9. Ibid., A 474/B 502.

10. Ibid., B 23.

11. Ibid., A 127. In order for understanding to impose its laws on nature and thus create nature, Kant claims that every concept requires an image that molds or fits the concept for application. In Kantian language such an image is called a "schema" or "schematized category." See A 140/ B 179-80; see also an essay written in 1786 titled "What is Orientation in Thinking," in which Kant writes the following: "However high we aim our concepts and however much we thereby abstract them from consequences, imaginal notions are always appended to them. Their proper function is to fit the concepts not otherwise derived from experience for empirical use." See Immanuel Kant, *Critique of Practical Reason and Other Writings in Moral Philosophy*, trans. and ed. Lewis White Beck (Chicago: University of Chicago Press, 1949), 293.

12. Ibid., A 126.

13. Ibid.

14. Ibid., B 148.

15. Ibid., B 27.

16. Ibid.

17. Kant, "Orientation in Thinking," 305. Compare *Critique of Pure Reason*, A 381-82. See also Immanuel Kant, *Groundwork of the Metaphysic of Morals*, trans. by H.J. Paton (New York: Harper & Row, 1964), 88. Here Kant states the most general formulation of the "categorical imperative": *"Act only on that maxim through which you can at the same time will that it should become a universal law."*

18. Ibid., 303-4.

19. Immanuel Kant, "What is Enlightenment?" in *Critique of Practical Reason and Other Writings in Moral Philosophy*, 286.

20. Ibid., 287.

21. Ibid., 290-91.

22. Kant, *Critique of Pure Reason*, A 805/B 833.

23. Ibid., A 850/B 879. The *"culture* of human reason" is the translation rendered by F. Max Muller. Muller's translation of Kant's original "die Vollendung aller Kultur" seems to capture the sense more directly than Kemp Smith's translation that omits the word *"culture."*

24. Immanuel Kant, "The Contest of Faculties," *Kant's Political Writings*, ed. Hans Reiss (Cambridge: Cambridge University Press, 1970), 186.

25. Kant, *Groundwork of the Metaphysic of Morals*, 100-101.

26. Kant, "The Contest of Faculties," 187.

27. Kant, "What is Enlightenment?," 292.

28. Kant, "Idea for a Universal History with a Cosmopolitan Purpose," *Kant's Political Writings*, 43.

29. Jean-Jacques Rousseau, "Discourse on the Sciences and Arts," *The First and Second Discourses*, ed. Roger D. Masters (New York: St. Martin's, 1964), 34.

30. Ibid., 47.

31. Ibid., 50.

32. Ibid., 39-40.

33. Ibid., 45.

34. Ibid., 44-45.

35. Ibid., 62-63.

36. See Chapter 2, "Fictional Visions of Science and Technology."

37. Ibid., 63.

38. David Hume, *The Natural History of Religion*, ed. H. E. Root (Stanford, Calif.: Stanford University Press, 1956), 76.

39. David Hume, *A Treatise of Human Nature: Being an Attempt to Introduce the Experimental Method of Reasoning into Moral Subjects*, ed. L. A. Selby-Bigge, 2d ed. (Oxford: Oxford University Press, 1978), 179.

40. David Hume, *Dialogues Concerning Natural Religion* [1779] (Indianaplois: Bobbs-Merrill, 1947), 159.

41. Hume, *A Treatise of Human Nature*, 1.

42. Ibid., 2.

43. Ibid., 3 and 4.

44. Ibid., 7-8.

45. Ibid., 8.

46. Ibid., 9. Hume notes that imagination is not "restrain'd to the same order and form with the original impressions."

47. Ibid., 11. It is important to remember that imagination is the other function or faculty of reason at work in the association of ideas. The principle regarding the function of memory is regarded in this context, as it is in Hume's discussion, as supported "by such a number of common and vulgar phaenomena" that it provides the most accessible means for grasping the systematizing function of reason in Hume's text. Hume thinks that, in light of these "common and vulgar phaenomena, we may spare ourselves the trouble of insisting on it further" (9). The "same evidence" holds for the "second principle, *of the liberty of imagination to transpose and change its ideas*" (10). The liberty of imagination will not appear strange once one considers, according to Hume, "that all our ideas are copy'd from our impressions, and that there are not any two impressions which are perfectly inseparable" (Ibid). But where imagination perceives a difference, it can produce a separation without difficulty.

48. One might consider Hume's *Treatise* as an explicit example of this systematizing function. Divided into three books, each book is divided into its own parts and sections, the *Treatise* presents three systems (labyrinths)—the systems "Of the Understanding," "Of the Passions," and "Of Morals." When taken as a whole, these systems constitute—by analogy—the system of the "human mind" (see Hume's introduction to the *Treatise*, especially xx-xxiii).

49. Ibid., 108.

50. Ibid.

51. Hume, *Dialogues Concerning Natural Religion*, 140.

52. Ibid.

53. David Hume, "Of the Original Contract," *Political Essays*, ed. by Charles W. Hendel (New York: Liberal Arts Press, 1953), 43.

54. David Hume, *An Enquiry Concerning the Principles of Morals*, (La Salle, Ill.: The Open Court, 1946), 1.

55. Hume, *A Treatise of Human Nature*, 149.

56. Ibid., 139.

57. Ibid., 221.

58. Ibid., 147. Compare bk. 1, pt. 3, sec. 11, "Of the Probability of Chances," 124-30; and sec. 14, "Of the Idea of Necessary Connexion," 155-72.

59. Ibid., 440.

60. Karl Marx and Frederick Engels, *The German Ideology*, in *Collected Works*, vol. 5, 44.

61. Ibid., 36.

62. Karl Marx, "Contribution to the Critique of Hegel's Philosophy of Law" [1844], in *Collected Works*, vol. 3, 176.

63. Ibid., 181.

64. The emphasis on the simultaneous articulation of the pessimistic and optimistic legacies in Marx's texts is a departure from an analysis that emphasizes the deterministic and voluntaristic elements of Marx's thought. For an account of this sort see, Langdon Winner, *Autonomous Technology*, 77-85. Contrary to the pessimism associated with technology as understood by Marx, there is an anlysis of the "bright vision" Marx provides. See Frederick Ferré, *Philosophy of Technology*, 54-57. In this context, see Bernard Gendron, *Technology and the Human Condition*, which provides a socialist critique of the utopian and dystopian views of technology.

65. Karl Marx, *Economic and Philosophic Manuscripts of 1844*, in *Collected Works*, Karl Marx and Frederick Engels (New York: International Publishers, 1975), vol. 3, 232.

66. Karl Marx and Frederick Engles, *The Holy Family or Critique of Critical Criticism* [1844], in *Collected Works*, vol. 4, 143.

67. Compare Karl Marx, *Economic and Philosophic Manuscripts of 1844*, in *Collected Works*, "[Estranged Labour]," vol. 3, 270-71. "We have proceeded from the premises of political economy. We have accepted its language and its laws. . . . Political economy starts with the fact of private property; it does not explain it to us. It expresses in general, abstract formulas the *material* process through which private property actually passes, and these formulas it then takes for *laws*. . . . it takes for granted what it is supposed to explain." By contrast, Marx claims to "proceed from an *actual* economic fact."

Marx's writings are replete with similar criticisms of other political economists. For example, see "The Poverty of Philosophy: Answer to the *Philosophy of Poverty* by M. Proudhon," in *Collected Works*, vol. 6, especially 111, 162-65, where the reference to Hegel's dialectics and "impersonal reason" has a prominent place.

68. Karl Marx and Frederick Engels, *The German Ideology*, in *Collected Works*, vol. 5, 231. "For the bourgeois it is all the easier to prove on the basis of his language the identity of commercial and individual, or even universal, human relations, as this language itself is a product of the bourgeoisie, and therefore both in actuality and in language the relations of buying and selling have been the basis of all others."

69. See Karl Marx, *Capital: A Critique of Political Economy* [1887] (Moscow: Progress Publishers, 1978), vol. 1, 29. "My dialectic method is not only different from the Hegelian, but is its direct opposite. . . . With me, . . . the ideal is nothing else than the material world reflected by the human mind, and translated into forms of thought. . . . With him [Hegel] it [the dialectic] is standing on its head. It must be turned right side up again, if you would discover the *rational kernel* within the mystical shell" [emphasis added].

70. Marx, "Contribution to the Critique of Hegel's Philosophy of Law," *Collected Works*, vol. 3, 182.

71. Karl Marx, *Economic and Philosophic Manuscripts of 1844*, in *Collected Works*, vol. 3, 304.

72. Marx and Engels, *The German Ideology*, in *Collected Works*, vol. 5, 36.

73. Marx, *Economic and Philosophic Manuscripts of 1844*, in *Collected Works*, vol. 3, 298.

74. Marx and Engels, *The German Ideology*, in *Collected Works*, vol. 5, 44.

75. Marx, *Capital*, vol. 1, 79.

76. Karl Marx, "On the Jewish Question," in *Collected Works*, vol. 3, 148. See G. W. F. Hegel, *The Phenomenology of Mind* [1807], trans. by J. B. Baillie (New York: Harper & Row, 1967), 136. "The series of shapes, which consciousness traverses on this road, is rather the detailed history of the process of training and educating consciousness itself up to the level of science."

77. Ibid.

78. Karl Marx, "The Eighteenth Brumaire of Louis Bonaparte," *Collected Works*, vol. 11, 103-4. "Men make their own history, but they do not make it just as they please; they do not make it under circumstances chosen by themselves, but under circumstances directly encountered, given and transmitted from the past. The tradition of all the dead generations weighs like a nightmare on the brain of the living. And just when they seem engaged in revolutionising themselves and things, in creating something that has never yet existed, precisely in such revolutionary crisis they anxiously conjure up the spirits of the past to their service and borrow from them names, battle-cries and costumes in order to present the new scene of world history in this time-honoured disguise and this borrowed language."

79. Ibid.

80. Karl Marx, *Grundrisse: Foundations of the Critique of Political Economy* [1857-58], trans. by Martin Nicolaus (New York: Vintage Books, 1973), 705.

81. Marx, *Capital*, vol. 1, 457.

82. Ibid., 364.

83. Dickens, *Hard Times*, 45.

84. Marx, *Capital*, vol. 1, 380. Emphasis added. In an analogous fashion, while noting the efficiency and productivity science and technology make available to society through the centralization of production (what is called today "economies of scale"), Marx points out that the accumulation of capital in the hands of a few capitalists bears the potential for increased exploitation and domination. Compare ibid., 588. "Everywhere the increased scale of industrial establishments is the starting-point for a more comprehensive organisation of the collective work of many, for a wider development of their material motive forces—in other words, for the progressive transformation of isolated processes of production, carried on by customary methods, into processes of production socially combined and *scientifically arranged*" [emphasis added]. See also, *Grundrisse*, 712.

85. Marx, *The German Ideology*, in *Collected Works*, vol. 5, 446-47.

86. Marx's recognition of the ambiguity discussed here is apparent in his revision and use of the "dialectic" method, not only in the scattered references to language.

87. Friedrich Nietzsche, *On The Genealogy of Morals* [1887], trans. Walter Kaufmann (New York: Vintage Books, 1967), essay 3, sec. 12.

88. Ibid.

89. Friedrich Nietzsche, *The Will to Power*, trans. Walter. Kaufmann and R. J. Hollingdale, ed. Walter Kaufmann (New York: Vintage Books, 1967), sec. 567.

90. Compare Nietzsche, *On the Genealogy of Morals*, essay 1, sec. 2. Nietzsche writes:

(The lordly right of giving names extends so far that one should allow oneself to conceive the origin of language itself as an expression of power on the part of the rulers: they say "this *is* this and this," they seal every thing and event with a sound and, as it were, take possession of it.)

Nietzsche echoes Marx's indictment of the use classical political economists and the bourgeois make of language as a tool for domination.

91. Ibid., sec. 618.

92. Ibid., secs. 618 and 466.

93. Ibid., sec. 632.

94. Nietzsche, *On the Genealogy of Morals*, 1st and 3d essays.

95. Friedrich Nietzsche, *Thus Spoke Zarathustra* [1883-84], trans. Walter Kaufmann (New York: Viking, 1966), 195.

96. Friedrich Nietzsche, "On Truth and Falsity in an Extra-Moral Sense,"[1873] in *Complete Works of Friedrich Nietzsche*, vol. 1, trans. Maximilian Mugge, ed. Oscar Levi (New York: Russell and Russell, 1964), 180.

97. Friedrich Nietzsche, *The Will to Power* , sec. 522.

98. Friedrich Nietzsche, *The Gay Science* [1882], trans. Walter Kaufmann (New York: Vintage Books, 1974), sec. 374.

99. Ibid., sec. 355.

100. Nietzsche, *The Will to Power*, sec. 606.

101. Ibid., sec. 477.

102. Friedrich Nietzsche, *The Use and Abuse of History* [1874], trans. Adrian Collins (Indianapolis: Bobbs-Merrill, 1957), 69-70.

103. Nietzsche, *On the Genealogy of Morals*, essay 3, sec. 24.

104. Friedrich Nietzsche, *The Birth of Tragedy Out of the Spirit of Music* [1872], trans. Walter Kaufmann (New York: Vintage Books, 1967), sec. 17.

105. Ibid., sec. 15.

106. Ibid, and sec. 21.

107. Nietzsche, *The Birth of Tragedy*, sec. 15. Compare *The Will to Power*, sec. 585 A. "A nihilist is a man who judges of the world as it is that it ought *not* to be, and of the world as it ought to be that it does not exist."

108. Nietzsche, *The Will to Power*, sec. 585 B.

109. Nietzsche, *The Birth of Tragedy*, "Attempt at a Self-Criticism," sec. 2.

The issue of authority has been treated, traditionally, as a question of replacement: one discourse is replaced with a more appropriate one, a discourse that fits the needs of achieving a particular goal. But reflecting on the concept of authority in this manner neglects to take into account the placement of the concept, the placement of any "authority" in the mesh of linguistic relations. Given this placement, the discussion of authority requires a shift in emphasis. It is no longer a question of physical coercion, brute strength, or of identifying specific qualities (e.g., through blood relations, through income and wealth, or through divine right) that endow an individual or an institution with power. Instead, it is a question of *rhetorical relations*: the relations displayed in the use of the word; the creation of specific linguistic devices, such as paradigms and models, based on and bound to the etymological/lexical relations that lend words their meaning and authority. However it is presented or deliberated, then, the question of authority is not merely one problem among others, nor is it the central problem to be addressed. It is, nonetheless, an issue integral to any deliberation or presentation concerned with science/technology. The authority of any text, scientific or other, is to be found in the endless play of metaphorical orders. The foundation of authority is always shifting from text to text and within each text. If there is any authority to be claimed, it is in the dissemination of the textual labyrinth.

In contrast to the perspective presented here, there are two other ways of analyzing the question of authority as it pertains to science and technology. One way to determine the authority or validity, that is the practical legitimacy of a particular scientific text, is by appeal to a specific set of criteria: a notion of objectivity, a concept of truth, a systematic ordering of all phenomena to mirror the order of nature, and methodological techniques that include internal coherence and consistency. Once this set of criteria has been satisfied, a text assumes a canonical status. It becomes another tool in the attempts to master nature. As such, the history of ideas would be read as a succession of canonical texts. This position will be referred to as the *classical-canonical* view of authority.

Another way to analyze authority is to introduce the notions of success and failure as they pertain to the operations of science and technology. This analysis forces a shift in emphasis from the theoretical to the practical. When scientific discoveries and technological innovations promise to benefit humanity, when they prove fruitful, they are granted privileged positions. However, when they fail to deliver on their promises, their legitimacy is subverted. This position will be called the *modern-assessment* view of authority.

Both of these accounts presume the possibility of authority; they presume that any-*thing* other than themselves, but including themselves, is subject to a universal rule. In other words, they presume any account can be legitimated. The criteria announced in each account will establish this possibility. In order for a narrative to become authoritative it must accept the rule of these determinations. Thus, according to either position—the classical-canonical view or the modern-assessment view—there is a definite ascent (and assent) to authority. Both presume an ideal, a foundation for authority. As such, each presumes a way out of the labyrinth.

The Classical-Canonical View of Authority

In whatever form it takes, authority is understood as rendering the final word on a particular subject or point of contention. Its word is absolute; it has become canonical. Authority presumes the resolution of debate, and as such the criteria that declare a settlement and determine the path to certainty. Furthermore, authority presumes obedience to its declarations. The authority of the Church, for example, is founded on the belief that the divine word is knowable, that it is communicated through institutional channels, and that it can be comprehended and obeyed by everyone. The divine word, as the first and final word, reveals the way to truth and salvation. Sometimes the authority of monarchs assumes a similar pretext: its absolute authority and sovereignty is grounded in a certain lineage of divine right. The monarch's power over life and death, assuming the right to terminate or preserve the life of anyone, is ordained by the institution of the monarchy itself, but often receives the blessings of religious authorities to supplement its reign. Yet, a monarchy's sovereignty is drawn as much from the power of its armies, its fortifications, and the economic control it exercises over its subjects as it is from its self-declarative posture.[1] Proclaiming themselves supreme authority, monarchs are the judges of all affairs; they render final judgments; their word is the final word in the settlement of a dispute. Like the authority of the Church, the authority of the monarch demands obedience, it has the power to extract duty from its citizens. Moreover, like the Church, the monarch promises truth and salvation. But there is a difference: the monarch promises peace, prosperity, and happiness in the here and now, not in an afterlife.

Rooted in the philosophic, scientific, and technological revolutions of the seventeenth century, the classical-canonical conception of authority (whether identified in seventeenth- or twentieth-century texts) stakes out the principles for evaluating its word as the first and final word. The sovereignty of science and the power of technology rest on a specific method of inquiry, not on divine revelation nor on the power money can buy. Despite working in a context where the boundaries separating philosophy/literature/science/technology are vague and duplicitous, each text endorsing the classical-canonical conception of authority aligns itself with a universal method of inquiry. To be sure, there are differences in how the methods of scientific inquiry are applied. But according to the classical-canonical conception of authority, there is a consensus traversing the differences; each method is organized by the belief in objectivity, truth, and the systematic ordering of all phenomena that mirrors or parallels the order of nature. There is, then, an appeal to some*thing* supratextual—*nature as such*. As already seen, Bacon's texts inspire the many directions this tradition pursues.

When Bacon introduces the four types of "idols" that "beset men's minds" or impede the development of human understanding and undermine access to certainty, he does so believing the idols are universally recognizable.[2] Bacon's categories (and one might imagine Kant's schema of categories in these terms as well)—concerning both the obstacles to knowledge and the means for overcoming these obstructions—are stated in general and "unbiased" terms. "The formation of ideas and axioms by true induction," according to Bacon, "is *no doubt the proper remedy* to be applied for the keeping off and clearing away of idols."[3]

Bacon does not appeal to the genius of a method that is his own or anyone else's. Instead he appeals to *the scientific method as such*: " . . . the course I propose for the discovery of sciences is such as leaves but little to the acuteness and strengths of wits, but places all wits and understandings nearly on a level."[4] Like the deductive method of inquiry, the scientific method proposed by Bacon claims universal application: "as the common logic, which governs by the syllogism, extends not only to natural but to all sciences, so does mine also, which proceeds by induction, embrace everything."[5] But Bacon's method introduces a difference; governed by induction it moves beyond what is given in already accepted axioms of logic. It anticipates what is not entailed in deduction. The legitimacy of Bacon's new method lies in the results it assures, and as such it usurps an authority typical of the impartial postures advocated by theological and political institutions.

What distinguishes the authority of scientific and technological inquiry from political and theological authority, according to Bacon, is the stress placed on the practices and techniques of laying a foundation for all inquiry—or for all possible knowledge. Knowledge is acquired through "active science," the activity of exploration; not through the acceptance and perpetuation of opinion and prejudice.[6] The method of science is prospective in just the sense that it is designed to bring about the "termination of infinite error . . . "[7] The hope of enlightenment provided by the scientific method, and initially articulated by Bacon, orients Kant's critiques,[8] as well as grounds the development of positivism from August Comte[9] through the Vienna Circle[10] to Habermas.[11]

The attempt to define what constitutes enlightenment, as Kant's and other enlightenment critiques show, is an attempt to establish foundations. Kant's critique of pure reason attempts, to use the words of Bacon, "to lay the foundation, not of any sect or doctrine, but of human utility and power."[12] In the tradition of the classical-canonical view of authority, the need for a foundation is seen as a need for stating the conditions and criteria for all possible knowledge and human action. Once a foundation for science is set in place, faithful acceptance of its laws and principles—its canons—is expected. It is an acceptance across the boundaries of other domains as well. In this respect, it involves a play of authority similar to that brought together in the political and theological domains.

The appeal to foundations is an appeal to a transcendental source— it is an appeal to some*thing* that lies beyond the immediate grasp of sensible perception. In theological terms the appeal is to a spiritual enlightenment that transcends religious rituals and institutional rules; in political terms the appeal is to natural rights and bureaucratic powers that transcend human experience, however veiled in extravagant ceremonies and costumes; and in scientific terms it is an appeal to a foundation or set of transcendental principles that transcend, but nevertheless are translated by, the development and use of highly technical and awe-inspiring instruments (telescopes, microscopes, and satellites), as well as the artificial languages of logic, mathematics, and computer science. Even though translated into and across different domains, it is the appeal to authority itself that is significant: by calling on powers allegedly external to their practices, all institutions declare the right to impose allegiance not only to themselves but also to their authority. The legitimacy of any appeal, of any authority thus depends on the context that it defines and within which it permits certain practices to take place.

The classical-canonical conception of authority has a certain aesthetic appeal for some. It is attractive: its lure is a set of criteria toward which one can turn—a set of apparently tangible rules and norms against which one can measure one's activities—either spiritually or scientifically, either personally or collectively. The criteria may change and can be revised for specific purposes; they may be challenged and repudiated. But, as Nietzsche points out in *On the Genealogy of Morals*, to those who would accept the canons of morality, theology, art, and science, adopting any set of criteria is better than having none. " . . . man would rather will *nothingness* than *not* will."[13] Canons are written and adopted to overcome disquietude. And, as the classical-canonical view of authority shows, order provides the means for achieving this end.[14] Moreover, the aesthetic appeal of this perspective is intertwined with the notion of replacement. Replacement connotes a sense of certainty and finality: it strengthens the desire for foundations. The replacement of one theory with another, of one system of order with an alternative, functions under the pretext of providing the final word. And the force of this word—of the canon—remains in the presentation of the world as it is—in its essence, isolated and fixed.

Indeed, the strength of the classical-canonical conception of authority is to be found in the ways it pictures the relations between nature, self, and society.[15] Nature—the world—is what it is because it is conditioned by a specific set of criteria used in scientific, philosophical, and theological discourse. It is stable, immutable, knowable (at least according to the criteria and principles that organize a specific narrative). The representation of reality in these terms is not conceived as the representation of a metaphorical system of order. It is designed to apprehend the world *as it is*, and to do so accurately and with certainty. As John Dewey notes, "The desire for intellectual or cognitive understanding had no meaning except as a means of obtaining greater security as to the issues of action."[16]

But what happens when the terms of understanding are changed? What happens when the world is conceived, as Dewey says, as "a scene of risk . . . uncertain, unstable, uncannily unstable" where even its "dangers are irregular" and "inconstant"?[17] Using Dewey's frame of reference, what was once considered the strength of the classical-canonical view is also its weakness: its foundation is no longer whole. "Any philosophy that in its quest for certainty ignores the reality of the uncertain in the ongoing processes of nature denies the conditions out of which it arises."[18] Thus, in a spirit that can be likened to a Nietzschean enlightenment, Dewey asserts that "the basic error of traditional theories of knowledge," and their attendant claims to authority, "resides in the isolation and fixation of some phase of the whole process of inquiry in resolving problematic situations."[19]

The consequences of the classical-canonical view of the authority of science and technology receive a more strident critique in the texts of Herbert Marcuse and Martin Heidegger. From the point of view articulated in Marcuse's *One-Dimensional Man*, science and technology impose a transformation on nature and humans that encroaches upon their "essence." There are no exceptions to the rule of science and technology; their operations are designed to institute a universal mode of control. Hence, the authority of science and technology invades every aspect of life. There is nothing that *is* or *will be* that is not conditioned or mediated by their rule.

The authority underlying the operations of science and technology, and thus the transformations these operations bring about, is given a priori—it is taken for granted. There could be no transformation of nature "into technical reality," no imposition of techniques of control, without accepting the authority of science and technology as such.[20] By turning nature, and all that nature involves, into instruments of control and organization, science and technology circumscribe "an entire culture; [they] project a historical totality—a 'world.' "[21] Marcuse underscores the force of the *"technological a priori"* when he cites Heidegger, who writes:

> "Modern man takes the entirety of Being as raw material for production and subjects the entirety of the object-world to the sweep and order of production (*Herstellen*). . . . the use of machinery and the production of machines is not technics itself but merely an adequate instrument for the realization (*Einrichtung*) of the essence of technics in its objective raw material."[22]

Even though Marcuse and Heidegger question the legitimacy of the canonical status assumed by and granted to science and technology, they recognize science and technology as means for "framing" the world. In other words, the use of science and technology is bound always to a particular context that is recast under their rule. The operations of science and technology are never neutral, but always destine the world (and, for Heidegger, Being) to appear in a specific way.[23] However, embedded within this critique is an ambivalence towards the authority of science and technology. Both Marcuse and Heidegger regard the canonical authority of science and technology as a significant means for altering how the world—nature, self, and society—is apprehended. The shift from a world view framed by theological principles and tenets to one framed by the methods of science is clearly one example of this point.

Such a shift in world views is a shift in authority: one authority supposedly replaces another. As a consequence of this shift, the practices of science and technology are (self-)validated, they become institutions. However, embedded within this (or any other) shift lie the possibilities of other shifts, transformations within and across the framework of any world view. Just as science and technology turn nature into an instrument, that is to say, something other than what it would be without their rule, science and technology can be turned against their "proper" canonical use—they become the instruments of critical, noninstrumental rationality. It is this condition, already examined in the texts of Hegel and Marx, that Marcuse and Heidegger celebrate. Whatever particular use is made of science and technology—the strategies for deploying their techniques within a particular cultural matrix—for Marcuse and Heidegger, it is always and only one use among a multiplicity of possible uses.[24]

It is clear, given the critiques of canonical authority registered by Dewey, Marcuse, and Heidegger, that this form of authority legitimates itself in terms of the comprehensive and efficient manner in which it represents the world. As Heidegger points out, technology validates the picture it presents of the world by ordering and concentrating its appearances, and thus by transforming the world into a "standing-reserve."[25] Pierre Duhem echoes a similar sentiment regarding the validity of the positions portrayed in/by scientific and technological texts. Although he focuses on the narratives of physical theory, what Duhem says about their validity and legitimacy can be extended to the narratives of science and technology in general: all narratives are motivated by the desire to "save the phenomena" in a representative fashion. But beyond merely saving the phenomena, the theories and hypotheses of physics, for example, are required to save "*all the phenomena* of the inanimate universe *together*." It is on this basis of uniformity and unity that narratives of this kind are generally regarded as classical-canonical systems. But, according to Duhem, they are nothing more nor less than "contrivances," that is to say fictional accounts used to isolate and fix the world.[26] Physics, like any other scientific theory for Duhem, is "*the increasingly better defined and more precise reflection of a metaphysics.*" Its "*sole justification*" is the belief "*in an order transcending*" its own narrative structures.[27]

Declaring knowledge objective and truth obtainable marks the canonical character of science and technology. But what is the basis, the authority for such a declaration? What difference in legitimacy is there between the declarations expressed in the texts of Bacon, Galileo, and Hobbes, for example, from the political declarations articulated in *The*

Declaration of Independence? Like the declarations of political and individual rights, the declarations of scientific and technological narratives function only within a specific, although not very clearly defined, context. And the context of these declarations is defined in the declarative act. What other foundation of authority is there?

Duhem ruminates over this question when he declares that *"a physical theory is free to choose any path it pleases provided that it avoids any logical contradiction; in particular, it is free not to take account of experimental facts."*[28] As long as a narrative avoids logical contradictions, it can claim the legitimacy of "science"; its claims to objectivity and truth can be validated. Scientific and technological narratives, then, are nothing more nor less than instruments. In a similar fashion, Karl Popper follows Duhem's critique. He argues that the conjectural power of theories and hypotheses is the only foundation for scientific/ technological authority. Any scientific narrative can be understood at best, according to Popper, as a "highly informative guess about the world," and thus remains open to refutation.[29] The only power that can be granted to science and technology, according to the critiques of Duhem and Popper, is the power of specific theories and practices as they demonstrate their falsifying value. Even though a theory works, even though a technique accomplishes its task or achieves the objective for which it is designed, its claim to authority is always provisional. And this radically provisional character of any authority, stressed through the critiques of the classical-canonical view of authority, introduces a subterfuge—the dispersal and displacement of authority by means of its own declarations.

The Modern-Assessment View of Authority

The classical-canonical view is concerned with authority only to the extent that it reflects the methods, goals, and foundations of scientific and technological inquiry and practice. One thematic thread appearing in each of the critiques of this view is the concern with the criteria that govern the operations of science and technology. Each critique shows an appreciation for the need to transform or revise the principles guiding any inquiry and practice, and in most cases encourages such a transformation. Every critique is the expression of a desire for its own governing principles or foundations. Even Dewey's "reconstruction" of philosophy and scientific method falls under this rubric as well.[30]

Even though the classical-canonical view of authority has been directly linked to the texts of Bacon and Kant, under its scope and appeal one can place the texts of Marx, Comte, and Habermas. Interwoven with the classical-canonical is a view of authority, and as such of science/technology, called the *modern-assessment view*. The modern-assessment view is interwoven with the classical-canonical in that it adheres to a conception of authority founded upon a recognizable set of criteria. Indeed, there is a difference in the specific criteria deployed from each perspective. For example, where the classical-canonical view is associated by some with the logical-empiricist tradition, the modern-assessment view has developed in concert with a "new age" or the "new synthesis" of science and technology.[31] Yet each perspective shares a belief in the need for articulating a foundation on which any new universal system can be erected. Without such a foundation or grounding, neither view can claim nor believes it possesses legitimacy.

Where the classical-canonical view is governed by a metaphysics of representation—a correspondence between narrative and reality—, the modern-assessment view is ruled by a metaphysics of evaluation and preservation. But the real differences between these two views is to be found elsewhere. The metaphysics of assessment presupposes the classical metaphysics: assessment is possible only when scientific/technological systems reflect the way the world is presumed to be. What distinguishes the modern from the classical view is a concern with preserving the "order" of nature and evaluating to what extent the progress of scientific/technological discoveries and inventions interfere with and alter that natural order. In this respect, the modern-assessment view emphasizes the practical aspects of the theoretical framework already set in place by the classical-canonical view of authority.

If there is more than one set of criteria and principles for legitimating the authority of science and technology, as the critics of the classical view suggest, what standard is to be used in choosing between them? If the standard is yet another theoretical principle or criterion, then one may fall into what appears to some to be an unavoidable infinite regress. In other words, if the task is to establish the final word, then the final word cannot refer to anything outside itself. To avoid such a perplexing set of circumstances, the modern-assessment view shifts attention to the practical domain. The ultimate appeal is to the positive application of any scientific/technological system. The success or failure of such a system is bound by its applicability. This is not to say that all assessment is of the risk/benefit kind, even though that is the most familiar and most frequently employed.[32] Where the risk/benefit analysis is guided by social and ethical norms, the more general assessment of science/technology is governed by an appeal to what is given—the facts. Frederick Ferré, for example, appeals to the authority of Alfred North Whitehead to substantiate this view when he cites, "The basis of all authority is the supremacy of fact over thought." According to Ferré:

> *Epistemological authority* must be grounded in what we find *given*, primitive to our reflection or elucidations. When (ideally) everything combines—when our *theories issue everywhere in successful methods* and their associated artifacts, and when our *practices are everywhere coherently and adequately interpreted by our theories*—that would be the *supreme authority* for intelligence.[33]

Ferré's comment captures the force of the modern-assessment view; it illustrates how contemporary critiques of science and technology presuppose the metaphysics of assessment. Even Jacques Ellul, who appeals to the culture of "techniques," understands scientific/ technological practices in classical terms. The "ultimate authority" for Ellul is not to be cast in terms of methodological success but rather in how the effects of techniques permeate the organization of society and culture. In this way, Ellul appeals to the cultural "facts" of technique(s). According to Ellul's assessment, the pervasiveness of techniques or technology in contemporary culture signifies its instrumental success. However, the prominence of technology is not bound to specific machines or tools. On the contrary, what makes technology appear as a threat—as already indicated by Marcuse and Heidegger—is its apparent autonomy. According to Ellul, "Technique has now become almost completely independent of the machine. . . . "[34] Further, "The human being is no longer in any sense the agent of choice. . . . He is a device for recording effects and results obtained by various techniques."[35] Because of this apparent autonomy, technique constructs a world— endowing it with order—in which it can operate. "It clarifies, arranges, and rationalizes; it does in the domain of the abstract what the machine did in the domain of labour. It is efficient and brings efficiency to everything."[36]

Ellul recommends no specific paths for dealing with the invasiveness of technique. His "final resolution" presents a critical account of technique as the means by which an individual "can express his ecstatic reactions in a way never before possible. He can express criticism of his culture, and even loathing. He is permitted to propose the maddest solutions."[37] Ellul's "final resolution" or answer to the question, "Is there any other way out [of the labyrinth of technique]?", maintains a legacy of ambiguity already apparent in many of the eighteenth-century enlightenments. Even though Ellul claims to be working with "realities and not with abstractions," and recognizes "the inevitability of the fact that technical difficulties demand technical solutions," he remains faithful to the "dreams of applying nontechnical remedies."[38] In effect, Ellul believes the spiritual can overcome the integration of the " 'man-machine' complex."

More pronounced and focused critiques that carry on the modern spirit of assessing the authority of science/technology are found in the texts of Langdon Winner and Joseph Agassi.[39] For both Winner and Agassi the possible "remedies," to use Ellul's term, are political in nature—assessment is a political issue, as are the solutions it recommends. According to Winner in *Autonomous Technology*, the task of assessment is not measuring and quantifying the impact of technological changes, but it is "an earnest effort to advance our understanding" about "the meaning of technology and the life of man."[40] In political terms, technology is assessed by Winner as a form of "legislation" influenced by a plurality of social and political perspectives.

While Winner recognizes the uncertainty and ambiguity characteristic of any remedy, he offers a set of maxims to indicate different passageways leading out of the technological labyrinth—"some specific principles to guide further technological construction."[41] These maxims are, in effect, elaborations of a "simple yet long overlooked principle: *Different ideas of social and political life entail different technologies for their realization.*"[42] Winner does not offer this principle as another theoretical or utopian solution; he criticizes Marcuse and Ellul for returning to a utopian orientation in their critiques. Instead, he understands the principle as signifying a way for grasping "the magnitude of what is to be overcome."[43] Moreover, it articulates a need to identify a variety of "appropriate" technologies, each of which can be assessed only within a particular context in terms of the specific purpose for which they are designed.[44]

So, the notion of an appropriate technology, where technology is understood in its "original" sense "as a means . . . like all other means available to us," is cast in terms of the following principles:

1. that technologies must be *"intellectually as well as physically accessible to those they are likely to affect"*;

2. *"that technologies be built with a high degree of flexibility and mutability"*; and

PHILLIPS MEMORIAL
LIBRARY
PROVIDENCE COLLEGE

3. *"that technologies be judged according to the degree of dependency they tend to foster, those creating a greater dependency being held inferior."*[45]

Winner, then, provides a set of criteria by which to assess the proper authority of science/technology. Winner's evaluation hearkens back to the classical view, in this respect, even though the appeal is to the factual and practical as opposed to the abstract and theoretical. For him the evaluation is based on the analysis of the extent to which technology can move towards establishing a quality of life where personal and political freedoms and rights are promoted. In this way, Winner reiterates the Kantian enlightenment optimism associated with the generation of scientific and technological inquiry.

Agassi's assessment of science/technology's authority is rendered along a political bias as well, a bias he identifies as "democratic." Believing "we need more technology, not less," Agassi focuses on the sort of techniques he thinks are the most appropriate means for controlling technology, that is means "required for the prevention of the catastrophes it may bring about . . . " For Agassi this is tantamount to a discourse on the technology of technologies.[46]

Like Winner and Ellul, Agassi looks to the social and political as the factual domains of technological interplay. Like other proponents of the modern-assessment view, Agassi's critique is guided by a specific principle, one that identifies the dangers and promises of "scientific technology." However, what distinguishes his critique from others within this view are the particular concepts employed in the articulation of his principle. For Agassi the overriding principle of authority assessment is the rational democratization of technology. Here Agassi fuses the classical and Popperian traditions: his appeal to rationality is classical, his appeal to open public debates in search of open-ended solutions is Popperian.

The problems of the modern world are more formidable than we ever faced. In this respect all opponents to technology are right. Yet a solution can only be found, if at all, by stepping technology up, not down, and by making it more rational, not less, as can be better achieved by democratization than by suppressing technocracy.[47]

Although Agassi and Winner are critical of what Paul Feyerabend calls the "theoretical authority" of science and technology, their texts or narratives still present a commitment to standards, that is to say a commitment to the need for a foundation without which no assessment is possible. In this respect, Winner and Agassi, like other critics of the classical view who assess authority according to the standards of the modern view, question the "social/political authority" of science and technology. Their critiques present ways—standards and values—for expanding the freedoms and liberties of individuals in the face of the invasive character of science/technology. But does the shift in focus from theoretical to social/political authority involve an appeal, for either Winner or Agassi, to the metaphysical or the instrumental value of democracy? Given the principles Winner recommends, one can conjecture that his appeal has a metaphysical twist. In an analogous manner, given Agassi's appeal to the ideals of democracy and rational discourse, one can conjecture that his appeal takes an instrumental or pragmatic turn.

One individual whose narratives have focused on the question of authority in social/political or "anthropological" terms, and who is considered a critic of both the classical-canonical and modern-assessment views, is Paul Feyerabend.[48] In its more strident moments, Feyerabend's critique of the classical-canonical conception of science/technology is an attack on the "theoretical authority" grounded in claims of universality, objectivity, rationality, and truth. According to Feyerabend, "it is clear, . . . that the idea of a fixed method, or of a fixed theory of rationality, rests on too naive a view of man and his social surroundings."[49] The example he cites in support of this claim deals with the relationship between the Copernican and Galilean theoretical systems. Feyerabend claims that "Galileo uses *propaganda*. He uses *psychological tricks* in addition to whatever intellectual reason he has to offer." In other words, "Galileo's utterances are indeed [rational] arguments in appearance only."[50] The science of nature as well as the nature of science, then, as it appears in Feyerabend's narrative, turns out to be the science of "natural interpretation"—in effect, the science of fabrication, *techné-logia*.

Feyerabend appreciates the fusion of the classical and modern views of authority and thus recommends the "balanced presentation" of their evidence. "The *theoretical* authority of science is much smaller than it is supposed to be. Its *social* authority, on the other hand, has by now become so overpowering *that political interference is necessary to restore a balanced development.*"[51] But does this political interference differ from the rules, policies, and procedures recommended by Winner and Agassi? According to Feyerabend's texts, herein lies a paradox that haunts any critique of the authority of science/technology. On the one hand, Feyerabend declares a principle: "anything goes." Not only does this principle herald the acceptance of methodological *pluralism*, but it exposes the *"fairy-tale of a special method"* aligned with scientific/technological inquiry.[52] On the other hand, what is significant about the principle Feyerabend articulates is its provision of a *foundation* on the basis of which to assess and evaluate authority. Feyerabend's principle supplies the framework for an anarchistic critique of knowledge and authority. Given what turns out to be the theoretical and practical supports for his assessment, Feyerabend advances the following claim:

> Facts alone are not strong enough for making us accept, or reject, scientific theories, the range they leave to thought is *too wide*; logic and methodology eliminate too much, they are *too narrow*. In between these two extremes lies the ever-changing domain of human ideas and wishes.[53]

Feyerabend's desire for "a multiplicity of ideas" can be achieved only through "the application of democratic procedures."[54] In spite of, and perhaps because of, his alleged positioning "outside" the scope of modern-assessment theory, Feyerabend can be aligned with Winner and Agassi through the lineage of the Kantian enlightenment spirit. He writes: "The attempt to increase liberty, to lead a full and rewarding life, and the corresponding attempt to discover the secrets of nature and of man entails, therefore, the rejection of all universal standards and of all rigid traditions."[55]

The classical-canonical and modern-assessment views of scientific and technological authority have been juxtaposed in terms of the apparent differences in the object of their respective accounts. On the one hand, the classical-canonical view accepts the theoretical principle of an already ordered universe or world. According to this view, once the secrets—the laws and regularities—of nature, self, and society are revealed, the possibility of mastering and controlling the world suddenly becomes real. On the other hand, the modern-assessment view is the perspective that nature, self, and society can be organized in accordance with specific practical maxims. The need for such practical maxims arises from the desire to control and master a complex set of circumstances that otherwise would not allow for realizing freedoms and rights. So, ostensibly both views assume a certain set of facts and values about the world, even though the particulars differ from one view to the other. Moreover, both views critically analyze the implications of accepting these facts, especially in terms of their joint articulation with the intervention of science/technology.

But, in effect, the modern-assessment view is the classical-canonical in disguise. Both attempt to retrieve—by their respective techniques—

the notion of a foundation. Both attempt to recover an order that exists outside their narrative techniques. And finally, both see themselves as having created a unique perspective from which to judge and criticize other selected accounts, a perspective not included in the scene they describe.

Given this account, should Feyerabend, for example, be included in the assemblage of these critiques of authority? After all, according to Feyerabend, there is no order, there are no "bare facts," there is no "fixed method" for uncovering the facts; there are only "interpretations of facts . . . already viewed in a certain way."[56] Yet, Feyerabend provides an account, a narrative based on a specific interpretive principle, of the ways in which the world has been ordered and should be ordered. Regardless of his insistence that his narrative is conditioned by anarchism, does he not present a theory that can be characterized as another "fairy-tale of a special method"? Is that not one of the effects it produces, that it can be characterized as such? Does he not attempt to arrest a futile debate regarding the science of nature and the nature of science? Does Feyerabend, in effect, foreclose on the drama of inquiry? Is his account yet another partial story running parallel to many others, all of which "mutually interlace and interfere at points" but none of which can be unified?[57]

The critique of scientific/technological authority, in either its classical-canonical or modern-assessment disguise, shares an analogous structure with the enlightenment critiques of religious and political institutional sovereignty. Like *The Declaration of Independence*, the pretext for the authority of science/technology is external to the narratives. Each narrative or text appeals to some regulative principle allegedly situated outside its structure and techniques, and each attempts to stand outside the context and conflicts it evaluates. But as is the case with the *Declaration*, the force and significance of each appeal, and thus the establishment of authority, is generated in and through a declarative act. The text announces its authority, for which it is the sole representative. In effect, the foundation of authority is not external to the narrative, nor is it comprehensible as a regulative principle; it remains internal to and constitutive of the context it fabricates. The narrative is its own foundation, and cannot be accounted for apart from the scene it describes.

Understanding the question of authority as it relates to science/technology, and understanding the labyrinth within which this question and science/technology appear, requires a retranslation of the question. The question is no longer a question of the theoretical or the practical aspects of authority isolated from one another. Instead, it is a question of use: the use made of the practical in the theoretical, the use made of the theoretical in the practical, and moreover the use made of their differences. The question of authority, then, is a question of the dissemination of differences across the cultural matrix fabricated through the interplay of scientific/technological, theoretical/practical, and etymological/textual domains. Authority cannot be located in any one domain; it is dispersed across the assemblage of domains. Such a dispersal is effected through the incessant generation of narrative accounts/fictions, each of which attempts to unify scattered themes and issues. In this respect, the dissemination of authority can be traced only within a labyrinth of different linguistic usage or techniques.

At best the classical-canonical or the modern-assessment view, and their attendant critiques, can provide partial or fragmented accounts of the question of authority. The exploration of this question has been traced through an ensemble of texts, the proper names of which designate certain thematic interests. This is not to say that the question of authority is central to any of these texts or narratives. Each outlines a passageway out of the labyrinth of narratives; each narrative account constitutes another path, another set of possibilities, or another assemblage of perspectives from which and with which to direct one's efforts within the labyrinth. To some extent, each narrative recognizes the role language plays in fabricating a critique or a system. Yet, in some way or another, each refuses or is unable to recognize itself as nothing more nor less than a fiction or a fairy tale or a legend. Is there any status other than that of a story or a fiction that a narrative can claim?

Of course Bacon, Hobbes, Galileo, Kant, and Marx are sensitive to this question. Each takes into account and gives an account of the authority language wields in the development of any systematic narrative. Yet they do not express the kind of ambiguity and ambivalence toward language that inform the narratives of Rousseau, Hume, and Nietzsche, whose accounts indicate a negative response to the question. It is this ambiguity and ambivalence that prompts a retranslation of the question of authority by Michel Foucault. Foucault is one of several contemporary thinkers who embraces this framework, attempting to redraw the limits of authority in terms of the use of language. For Foucault, language is not only a technical tool or instrument for the representation of reality or the world. Instead, language is the medium in which rules and codes determine the category of representation itself.

According to Foucault, the examination of scientific knowledge does not proceed "from the point of view of the individuals who are speaking, nor from the point of view of the formal structures of what they are saying, but from the point of view of the rules that come into play in the very existence of such discourse. . . . "[58] As such, scientific discourse must be approached from a multiplicity of levels and methods simultaneously because of its apparent complexity. At each level, deployment of a particular method groups, isolates, analyzes, matches, and pigeonholes "concrete contents." However, as Foucault notes, even with the use of such techniques "there is nothing more tentative, nothing more empirical (superficially, at least) than the process of establishing an order among things; nothing that demands a sharper eye or a surer, better-articulated language. . . . "[59] Order in this context is another designation for "the grid created by a glance, an examination, a language; and it is only in the blank spaces of this grid that order manifests itself in depth as though already there, waiting in silence for the moment of its expression."[60]

Order is established through the "fundamental codes of a culture" that govern "its language, its schemas of perception, its exchanges, its techniques, its values, the hierarchy of its practices."[61] The codes of a culture, those that frame the grid or what has been called a cultural matrix, attempt to secure a "direct" relation with nature, self, and society through representation. The "proper names" of those objects to be represented are "mere" artifices; they give us, as Foucault says, "a finger to point with, in other words, to pass surreptitiously from the space where one speaks to the space where one looks; in other words, to fold one over the other as though they were equivalent."[62] However, that finger or point of reference is an image "that has just escaped from the open pages of a book."[63] Thus, the point of reference itself—the world, nature, reality, or the self—is nothing more nor less than a textual point, its "whole being is nothing but language, text, printed pages, stories that have already been written down."[64] In this respect, Foucault abandons the canons of representation underwriting the classical and modern assessments of authority. Scientific discourses, like all "written texts," are "extravagant romances." They are "quite literally, unparalleled: no one [thing] in the world ever did resemble" their characters; "their timeless language remains suspended, unfulfilled by any similitude; they could all be burned in their entirety and the form of the world would not be changed."[65]

Foucault's attempts to rethink the concept of authority in terms of power relations take on many forms. His examinations of the historical, methodological, and cultural interplay of economics, politics, medicine, art, and sexuality provide the bases for these attempts. Each examination shares a concern with locating "the forms of power, the channels it takes, and the discourses it permeates. . . . "[66] Though Foucault does not conceive power in physical or coercive terms, he sees it as "the overall effect that emerges from all of these mobilities, the concatenation that rests on each of them and seeks in turn to arrest their movement."[67] Fixing or isolating a discursive moment or historical epoch is not achieved solely in the theoretical or practical domain. Power is, as Foucault points out, "everywhere . . . because it comes from everywhere, . . . insofar as it is permanent, repetitious, inert, and self-reproducing."[68]

If power or authority is everywhere as Foucault claims, it is because of its duplicity of appearance. Power/authority/knowledge appears at once to be permanent and self-reproducing; it appears at once to be repetitious and different; it appears at once to be inert and active. To account for these differences, for this duplicity, Foucault focuses on the relations constitutive of what is called by the name *power*. On these grounds, according to Foucault, the descriptive accounts of certain "terminal forms" of authority, that is power relations that appear fixed and static, are themselves involved in the institution of these supposed fixed and uniform power relations. Thus, discursive relations constitute power relations or authority in recognizable, that is to say as ordered, forms.

Discursive relations, then, are conditioned by the differences that regulate the kinds of discourse involved in any narrative. As already indicated, narratives incorporate relations of etymology, grammar, and syntax, as well as history, method, and system. But none of the discursive differences in levels or in kind constitute what Lyotard calls a "*genre*"—an independent category in the Kantian sense or a "grand-narrative."[69] Like any word, any phrase, or any sentence or text, a kind of discourse is always intertwined with other discourses at a multiplicity of levels. "As soon as there is one sentence," one text, one discourse, "there are several. The one comes up along with the multiple."[70] Incorporated into an assemblage of narratives, each discourse is a "presentation," as Lyotard notes: it presents the assemblage and, as such, an "unpresentable" "universe." Whatever form it takes, the presentation involves an ontological, a metaphysical, and an epistemological " 'there is.' "[71] As such, a text—a narrative, a discourse, a fiction—is at once a point of convergence or condensation—where themes and issues are brought together—and a point of rupture—where boundaries are ignored and redrawn.

Meta-Narratives and Dissemination

Whether authority is discussed in terms of the classical-canonical conception, the modern-assessment view, or in terms of retranslating the theoretical into the practical and both the theoretical and practical in terms of the linguistic, it is linked to the practices of replacement. Narrative accounts, discourses, and texts—including their various techniques and devices—arrange a set of power relations, in a systematic fashion, to promote one fiction over another. In their respective and subtle ways, narratives intend to replace one another—their fiction is designed to subvert the power of others. Moreover, this subversion is a move towards dominating the form and style of discourse, the field of discourse, as well as legitimating itself and the field it fabricates. In order to assume this posture of authority, a narrative must have ascended to a Meta-state, what Lyotard calls the status of a grand-narrative.

Presumed at every moment of replacement is the ultimate foundation on which knowledge and authority can be established. Replacement, then, becomes the means for attaining the final word; this is certainly the tendency of the narratives that compose the classical-canonical and modern-assessment views. But, if the foundation of authority is to be plotted within texts always involved in the scene they describe and criticize, and as such attempt to legitimate themselves, "there is" no replacement, only practices of displacement. And according to these practices, all texts, all fictions, all levels of representation remain points of reference for each other. These points of reference become the pegs on which metaphorical systems are constructed. In other words, there are only narratives, legends and legacies, texts and discourses—the unraveling of a labyrinth in which they are paths and passageways.

There is a sense in which any narrative or text becomes a *meta*narrative but never a *Meta*–narrative.[72] Written with a capital *M*, a Meta-narrative would present a privileged perspective *outside* the labyrinth of narratives, fictions, texts. However, since such a privileged stance can never be fabricated, one can expect only to present narratives, that is narratives written *within* the labyrinth. Even though some of these narratives would present a privileged perspective of the etymological/textual, theoretical/practical, and scientific/technological domains (claiming the status of Narratives), by their very mode of inscription they remain metanarratives. Written in lowercase, *meta*narratives are no longer fashioned in a distinct manner. They cross the universal boundaries separating these domains; their descriptions and critiques are bound always to the labyrinth, and mark their involvement in the incessant generation of the labyrinth.

Written with a lowercase *m*, metanarratives diffuse the reductionist orientation of Meta-narratives. Instead of providing a set of criteria, an ultimate foundation, a final word, metanarratives issue no-thing. There is only the endless issue of interpretations. In other words, as soon as there is one interpretation there is a multiplicity of interpretations. Each narrative is an interpretation *of* or *about* another narrative; each account generates other interpretive accounts. In effect, each narrative goes beyond itself and becomes a metanarrative. Only with the possibility of *other* interpretations, then, is there the possibility of one interpretation. In this sense, there is a labyrinth of interpretations or narratives or fictions.[73]

As such, metanarratives disseminate authority: they proliferate and perpetuate interpretive practices. This performance is all "there is." There can be no "centers" in which power is concentrated nor figures in which authority resides. There are only techniques—the deployment of narratives for specific purposes that define the context of their use.

Notes

1. For another version of this point, see Michel Foucault, *The History of Sexuality, Volume I: An Introduction*, trans. Robert Hurley (New York: Vintage Books, 1980), 136ff.

2. Francis Bacon, *The New Organon*, aphorism 39, 47.

3. Ibid., aphorism 40, 48. Emphasis added.

4. Ibid., aphorism 61, 58.

5. Ibid., aphorism 127, 115-116.

6. Francis Bacon, Preface to *The Great Instauration, The New Organon and Related Writings*, 16.

7. Ibid.

8. Kant acknowledges his debt to Bacon. The epigraph to the *Critique of Pure Reason* (B ii) is taken from the preface to Bacon's *The Great Instauration*.

9. Compare Auguste Comte, *The Second System: Système de Politique Positive* [1851-54], in *Auguste Comte and Positivism: The Essential Writings*, ed. Gertrude Lenzer (New York: Harper & Row, 1975), 331. "The abstract study of nature is therefore all that is absolutely indispensable for the establishment of unity in human life. It serves as the foundation of all wise action—as the *philosophia prima*, the necessity of which in the normal state of humanity was dimly foreseen by Bacon."

10. The notion of "protocol sentences," as the fundamental units from which to construct a whole edifice, is most pronounced in Otto Neurath, but is a pervasive theme in the writings of all members of the Vienna Circle, especially Rudolf Carnap.

11. Habermas readily accepts the legacy of the classical-canonical view of authority when he claims that the "Enlightenment project" is incomplete. Compare Jürgen Habermas, "Modernity—An Incomplete Project," in *The Anti-Aesthetic: Essays on Postmod-*

ern Culture, ed. Hal Foster (Port Townsend, Wash.: Bay Press, 1983), 3-15; "Philosophy as Stand-In and Interpreter," in *After Philosophy: End or Transformation?*, ed. Kenneth Baynes, James Bohman, and Thomas McCarthy (Cambridge: MIT Press, 1987), 296-315; and *The Theory of Communicative Action, Volume I: Reason and the Rationalization of Society*, trans. Thomas McCarthy (Boston: Beacon, 1981), see especially 1-43 and 102-43.

12. Kant, *Critique of Pure Reason*, Bii.

13. Nietzsche, *On the Genealogy of Morals*, essay 3, sec. 28.

14. Ernest Gellner, *Legitimation of Belief* (Cambridge: Cambridge University Press, 1974), 125. "Chaos is not an option. No society, no culture or language, either does or can exist, which operates on the assumption of a chaotic nature, of a world not amenable to conceptual order."

15. Compare John Dewey, *Experience and Nature* [1929] (New York: Dover, 1958), 128.

16. John Dewey, *The Quest for Certainty: A Study of the Relation of Knowledge and Action* [1929] (New York: Capricorn Books, 1960), 38.

17. Ibid., 41.

18. Ibid., 244.

19. Ibid., 188.

20. Herbert Marcuse, *One-Dimensional Man: Studies in the Ideology of Advanced Industrial Society* (Boston: Beacon, 1964), 154.

21. Ibid.

22. Ibid., 153-54. Compare Martin Heidegger, *Holzwege* (Frankfurt: Klostermann, 1950), 22 and 29.

23. Martin Heidegger, "The Question Concerning Technology," [1954] in *The Question Concerning Technology and Other Essays*, trans. William Lovitt (New York: Harper & Row, 1977), 19-27.

24. Compare Ibid., 23-27. See also Marcuse, *One-Dimensional Man*, 230-32 and "Industrialization and Capitalism in the Work of Max Weber," in *Negations: Essays in Critical Theory* (Boston: Beacon, 1968), 215.

25. Heidegger, "The Question Concerning Technology," 19.

26. Pierre Duhem, *To Save the Phenomena: An Essay on the Idea of Physical Theory from Plato to Galileo* [1908], trans. Edmund Dolan and Chaninah Maschler (Chicago: University of Chicago Press, 1969), 117.

27. Pierre Duhem, *The Aim and Structure of Physical Theory* [1906], trans. Philip P. Wiener (Princeton: Princeton University Press, 1954), 335.

28. Ibid., 206.

29. Karl R. Popper, *Conjectures and Refutations: The Growth of Scientific Knowledge* (New York: Harper & Row, 1963), 115. Compare Ian Hacking, *Representing and Intervening: Introductory Topics in The Philosophy of Natural Science* (Cambridge: Cambridge University Press, 1983), 145. "In physics there is no final truth of the matter, only a barrage of more or less instructive representations." In saying this, Hacking acknowledges his debt to Duhem, Dewey, and Popper. See 130, 143, and 145.

30. Compare John Dewey, *Reconstruction in Philosophy* [1920] (Boston: Beacon, 1948), ixff. Here Dewey articulates what he calls the reconstruction of "intelligence" or philosophical inquiry—"the method of observation, theory as hypothesis, and experimental test."

31. Anne L. Hiskes and Richard P. Hiskes, *Science, Technology, and Policy Decisions* (Boulder: Westview Press, 1986), see especially chap. 1, "Science and Technology: Public Image and Public Policy."

32. Compare K.S. Shrader-Frechette, *Nuclear Power and Public Policy: The Social and Ethical Problems of Fission Technology* (Dordrecht: Reidel, 1980), see especially 156-58; "The Real Risks of Risk-Cost-Benefit Analysis," in *Technology and Responsibility*, ed. Paul T. Durbin (Dordrecht: Reidel, 1987), 343-57. See also: Maurice N. Richter, Jr., *Technology and Social Complexity* (Albany: State University of New York Press, 1982), Chap. 7, "Impacts of Technology," 84-97; Teich, *Technology and the Future*, see sec. 2, "Forecasting, Assessing, and Controlling Technology," 133-258; A. Pablo Iannone, ed., *Contemporary Moral Controversies in Technology* (New York: Oxford University Press, 1987), especially sec. 2, "Moral Controversies in Technology Assessment," 47-75; and Rudi Volti, *Society and Technological Change* (New York: St. Martin's, 1988), pt. 1, chap. 2, "Winners and Losers: The Differential Effects of Technological Change," 16-31.

33. Ferré, *Philosophy of Technology*, 51. Emphasis added.

34. Ellul, *The Technological Society*, 4.

35. Ibid., 80.

36. Ibid., 5.

37. Ibid., 424.

38. Ibid., 414.

39. Other texts exemplifying a critique that articulates a principle for managing the authority of science/technology include: Gendron, *Technology and the Human Condition*, where a socialist view overrides a concern with assessment by announcing the "prospects for emancipation" (pt. 3); and Borgmann, *Technology and the Character of Contemporary Life*, where technological reform has as its aim "the recovery of the promise of technology," that is to say a recovery of the authenticity of "life" through political and economic freedom and responsibility (see 101-13 and 226-49).

40. Winner, *Autonomous Technology*, 7. See also the reference to "forms of life" and "mode of life" in *The Whale and the Reactor*, chap. 1, "Technologies as Forms of Life," 3-18.

41. Ibid., 326.

42. Ibid., 325.

43. Ibid., 326.

44. Winner, *The Whale and the Reactor*, chap. 4 "Building the Better Mousetrap," 61-84. Compare E.F. Schumacher, *Small is Beautiful: Economics as if People Mattered* (New York: Harper & Row, 1973). Schumacher discusses the notion of appropriate technology in terms of economic mechanisms and tools implemented in less developed countries or developing countries.

45. Winner, *Autonomous Technology*, 326-27.

46. Agassi, *Technology*, 1.

47. Ibid., 258.

48. Paul Feyerabend, *Against Method: Outline of an anarchistic theory of knowledge* (London: NLB/Verso, 1975/78), 252. In this regard, Feyerabend claims "my argument presupposes, of course, that the anthropological method is the correct method for studying the structure of science (and, for that matter, of any other form of life)." Compare Richard J. Bernstein, *Beyond Objectivism: Science, Hermeneutics, and Praxis* (Philadelphia: University of Pennsylvania Press, 1983) for a protracted critique of the anarchism promoted by Feyerabend and that Bernstein associates with the writings of others, e.g., Kuhn, Rorty, and Derrida.

49. Ibid., 27. Compare Thomas S. Kuhn, *The Structure of Scientific Revolutions*, 2d ed. (Chicago: University of Chicago Press, 1970). Kuhn is, of course, the most notorious critic of theoretical authority of the scientific establishment. His critique of authority focuses on the internal sociological dynamics of the scientific community. "As the source

of authority, I have in mind principally textbooks of science together with both the popularizations and the philosophical works modeled on them" (136).

Compared to recent feminist critiques of the social and political forces influencing the authority of science/technology, Kuhn's account, and perhaps Feyerabend's, are "commonplace." For several descriptions of this situation see: Evelyn Fox Keller, *Reflections on Gender and Science* (New Haven: Yale University Press, 1985), 3ff.; Sandra Harding, ed., *Feminism and Methodology* (Bloomington: Indiana University Press, 1987); and Rothschild, ed., *Machina ex Dea: Feminist Perspectives on Technology* (New York: Pergamon, 1983).

    50. Ibid., 81.

    51. Ibid., 216.

    52. Compare Ibid., 27, 52, and 303.

    53. Ibid., 303.

    54. Ibid.

    55. Ibid., 20.

    56. Ibid., 19.

    57. James, *Pragmatism*, 71.

    58. Foucault, *The Order of Things*, xiv; compare Foucault, *The History of Sexuality, Volume I: An Introduction*, 11, where Foucault assumes a position on this point of "method" apparently contrary to that expressed in *The Order of Things*. "The central issue then . . . [is] to account for the fact that it [sex] is spoken about, to discover who does the speaking, the positions and viewpoints from which they speak, the institutions which prompt people to speak about it and which store and distribute the things that are said." But these differences can be accounted for in the following way. According to Foucault, again taken from *The Order of Things*, "Discourse in general, and scientific discourse in particular, is so complex a reality that we not only can, but should, approach it at different levels and with different methods" (xiv).

    59. Ibid., xix.

    60. Ibid., xx.

    61. Ibid.

    62. Ibid., 9.

    63. Ibid., 46. See also the discussion of the relationship between the term *pipe*, the pictorial representation of a pipe, and the physical object known as a pipe. Michel Foucault, *This is not a Pipe*, trans. and ed., James Harkness (Berkeley: University of California Press, 1983).

    64. Ibid.

    65. Ibid.

    66. Foucault, *The History of Sexuality*, 11.

    67. Ibid., 93.

    68. Ibid.

    69. Lyotard, *Le Différend* par. 190.

    70. Ibid., par. 103; see also Jean-François Lyotard, "Presentations," in *Philosophy in France Today*, ed. Alan Montifiore, trans. Kathleen McLaughlin (Cambridge: Cambridge University Press, 1983), 129.

    71. Ibid., par. 111.

    72. The distinction between *metanarratives* and *Metanarratives* parallels the distinction between *enlightenment* and *Enlightenment* that was explored in chap. 3. Compare Richard Rorty, *Consequences of Pragmatism* (Minneapolis: University of Minnesota Press, 1982), xl-xliv. Here Rorty distinguishes between *philosophy*, in terms of a post-

Philosophical culture, and *Philosophy* as the designation for the history of ideas as a systematic tradition.

73. The use of the term interpretation here follows a Nietzschean legacy. Nietzsche, Derrida, Lyotard, and others distinguish *interpretation* as an "activity" from the theoretical framework of *philosophical hermeneutics* as it has been articualted, for example, by Hans-Georg Gadamer, *Truth and Method* [1960], trans. Garrett Barden and John Cumming (New York: Seabury Press, 1975); see especially pt. 3, "The Ontological Shift of Hermeneutics Guided by Language," 345-498. Compare Bernstein, who in *Beyond Objectivism and Relativism* fuses the Gadamerian hermeneutical framework with a Habermasian critical foundation. There is a marked difference between what Bernstein calls "the hermeneutical dimension of science" and the labyrinth of interpretations at work in the current discussion.

CHAPTER 5
CONSEQUENCES OF DISSEMINATION
Narrative Recollections and
the Languages of Pedagogy

Post-Philosophy: The Contemporary-Linguistic View of Authority

All narratives, whether they are cast as "philosophical" inquiries, "scientific" treatises, or "literary" texts, are inextricably bound to a labyrinth of interpretations, that is a web of other fictions, other narratives. Informed by another set of concerns, or what might be thought of as another "Philosophical tradition," Richard Rorty recognizes this intractable condition by commenting on what he identifies as the "post-Philosophical culture."[1] According to Rorty, the condition that marks narrative accounts presents what he calls "a bedrock metaphilosophical issue." Rorty asks if one can "ever appeal to nonlinguistic knowledge in philosophical argument?"[2] Another way of posing this question is to ask whether or not one can appeal to any*thing outside* the labyrinths in which any interpretation is fabricated and must be comprehended?

Posing this question, Rorty comments on the possibility of securing a "foundation," a transcendental criterion underlying all knowledge and, thus, all philosophical arguments. How one responds to this question, according to Rorty, will determine the orientation of one's future philosophical inquiry. According to Rorty, two ostensible responses are possible and can be framed in terms of the confrontation between the "pragmatist" and the "intuitive realist." As far as Rorty is concerned, it does not matter which response is given because the choice is always between finding *"natural starting points which are distinct from cultural traditions"* or comparing and contrasting *"cultural traditions."*[3]

According to the vocabulary of traditional Philosophical discourse, the distinction Rorty articulates can be understood in terms of the classical opposition between *relativism* and *absolutism*. Stated briefly, it can be said that while the relativist position acknowledges the differences and legitimacy of multiple cultural traditions, the absolutist position appeals to a point of reference underlying and independent of all traditions. Rorty obviates this traditional bifurcation by introducing a different vocabulary: for Rorty the choice is no longer between absolutism and relativism but between Philosophy and a post-Philosophical culture. Philosophy is the tradition that perpetuates schemas of classification and the binary opposition of concepts. Overcoming the weaknesses or insufficiencies of one classificatory framework is achieved by replacing it with another, supposedly more comprehensive, one. In this way, the tradition of Philosophy seeks an algorithm, that is "criteria to which all sides must appeal" in order to engage in argumentation and to resolve disputes of substance. However, it should be noted that in the idiom of Rorty's commentary, the post-Philosophical culture *replaces* the central role assigned to classical conceptual oppositions by focusing on the differences constituting the vocabularies used in any particular context. According to Rorty, in a post-Philosophical culture, then, "criteria would be seen as the pragmatist sees them—as temporary resting-places constructed for specific utilitarian ends."[4]

In spite of his claim that "it does not matter which description [narrative or vocabulary] one uses," Rorty's account is torn between two extremes. On the one hand, even if Rorty's criterion is *pragmatic* in character rather than *positivistic* (or *verificationist*) in principle, he proffers another criterion for preferring the post-Philosophical to the Philosophical culture. The criterion is defined by its use. The demarcation of the linguistic and the nonlinguistic vocabularies of knowledge, the distinction between *systematic* and *edifying* discourses, or the foundationalist and nonfoundationalist, is itself a function of linguistic differentiation or commentary.[5] Thus, as a pragmatic criterion Rorty assumes it can supply a temporary resting place from which other commentaries can be generated, nothing more nor nothing less.

But on the other hand, by way of declaration, that is by the pragmatics of the situation defined by Rorty, he claims the distinction to be central to any philosophical inquiry. By identifying the opposition of philosophical cultures as "a bedrock metaphilosophical issue," Rorty aligns

himself with traditional Philosophical absolutism, reducing or translating the complexity of differences to a single binary opposition. It
seems that the "pragmatic" character of all criteria or rules becomes
the new "natural" point of departure in the post-Philosophical culture.
Of course, Rorty would deny that his narrative provides any such
proscription.[6] But given the essential role assigned to the bifurcation in
Rorty's narrative, how is it that his account supplies nothing more nor
less than (a) temporary resting place(s)?

Rorty's commentary on the bifurcation of philosophical vocabularies
functions at a Meta-narrative level, even though it attempts to avoid in-
stituting itself as such. The simple determination of different philosophical cultures, traditions, or vocabularies is framed within the labyrinth of interpretations and narratives. And yet, in order to recognize
the differences, Rorty's account must assume a position outside the
labyrinth; it must assume the neutral position of a mediator. In order to
redefine philosophical vocabularies, it must ascend to the level of a
Meta-narrative, which involves an appeal to nonlinguistic knowledge.[7]
Despite statements to the contrary, Rorty articulates "a set of rules
which tell us how rational agreement can be reached on what would
settle the issue on every point where statements seem to conflict."[8] In
other words, this set of rules allows the construction of a common
ground and orders the linguistic boundaries within which understanding—the commensurability of "ideal situations"—is possible.
Rorty, then, presents a commensurable point of reference from which
some form of agreement and conversation is anticipated. In order to
converse with Rorty on this issue, agreement is required. He narrates
the opposition of Philosophy and post-Philosophy in such a way as to
present a view that comprehends the *real* issues or differences at stake.
Thus, Rorty's narrative requires consensus, even if it is an agreement to
recognize differences.

Rorty's notion of commensurability differs from the notion of commensurability of vocabularies or linguistic usage where "the same
meaning" is assigned to terms.[9] By attempting to determine certain
points of agreement in a systematic fashion, for the purpose of interpreting differences between vocabularies and understanding those differences in pragmatic terms, Rorty's notion can be called *discursive*
commensurability. He recasts the legacies and enlightenments of the
Anglo-American analytic tradition in such a way that his account fashions what may be called a *contemporary-linguistic* view of authority. It
is a view constituted vis-à-vis Quine's notion of the indeterminacy of
translation and the incommensurability of theories as explored by
Kuhn and Feyerabend.

The contemporary-linguistic view of authority forces a shift in focus from the relation of the theoretical and the practical to linguistic usage. The importance of this shift lies in the recognition that the constitution and determination of any theoretical/practical domain is a function of language. But, does this shift herald a change in how the authority of particular narratives and institutions is or should be conceived? Does emphasizing linguistic usage transform the question of authority as it is manifested according to the classical-canonical and modern-assessment views? Or, can the authority of linguistic usage be seen, like the modern-assessment view, as another disguise of the classical-canonical view?[10]

One reason for identifying the modern-assessment as a disguise of the classical-canonical view is that both views advance a set of standards or criteria according to which a preference can be made regarding competing and alternative theories or models, systems, or interpretations. In a similar fashion, the contemporary-linguistic view of authority upholds a set of criteria for choosing between different visions of culture—philosophical, scientific, literary. That is to say, "rules" are laid out in advance of having to decide; they articulate and maintain lines of demarcation before the choice between differences becomes an issue. One can imagine the deployment of such criteria in drawing a distinction between *meaningful* and *meaningless* linguistic expressions (as exemplified by the practices of the Vienna Circle), between less or more *appropriate* modes of discourse (as presented in *ordinary-language* debates), or between interpretations that are more or less *comprehensive* (as expressed in the debates of philosophical hermeneutics).

In each case, or according to each view of authority, the only means by which progress can be made, in terms of overcoming the limitations of a prevailing narrative, is to move outside the labyrinth of narratives and interpretations—that is, to reveal and disclose a set of intuitive criteria or first principles. Though the contemporary-linguistic view of authority focuses on the differences engendered through language use, it still attempts to provide an order; it still attempts to analyze, systematize, and organize narratives in terms of opposition and hierarchy. As opposed to an account of disseminated narrative authority, the contemporary-linguistic view maintains a faith in an authority that is discernable and concentrated in the declarations of specific post-Philosophic texts.

Among the many and varied treatments of language that have been advanced in contemporary philosophical, scientific, and literary discourses, the theme of the indeterminacy of translation, as developed by Willard V.O. Quine, underlies Rorty's perspective and is an integral element of the contemporary-linguistic view of authority. According to Quine, the indeterminacy of translation is not intended to characterize the whole linguistic maze. In general, translation is possible between "kindred" languages and between "unrelated" languages due to the "resemblance of cognate word forms" and "traditional equations that have evolved in step with a shared culture" respectively.[11] By referring to the resemblance of cognates, Quine invokes both Neurath's and Carnap's models of the logical and empirical reduction of natural languages.[12] In effect, then, by referring to traditional equations of a shared culture, Quine invokes the legacies of positivism and behaviorism.[13]

However, the indeterminacy of translation is pertinent only to what Quine calls "*radical* translation, i.e., translation of the language of a hitherto untouched people."[14] In other words, the indeterminacy of translation deals with the incompatibility of cultures—philosophical or other—where the incompatibility is defined in terms of languages that do not correspond, that is languages that are neither kindred nor unrelated according to Quine's determinations. "Manuals for translating one language into another can be set up in divergent ways, all compatible with the totality of speech dispositions, yet incompatible with one another."[15] The incompatibility becomes more radical as the direct links between sentences and nonverbal stimulations weaken. Here Quine abandons the simple notion of translation that presupposes a one-to-one correspondence between the words and sentences of one language and those in another.

Linguistic use, according to Quine, forms and is formed within a theoretical framework: "We may well have begun then to wonder whether meanings even of whole sentences (let alone shorter expressions) could reasonably be talked of at all, except relative to other sentences of an inclusive theory."[16] To talk about the "truth" of sentences within the parameters of an inclusive theory is not precluded because "one is always working within some comfortably inclusive theory, however tentative." However relative or provisional the notion of truth, it is "overtly relative to language. . . ."[17]

Complete and radical translation from one language to another, from one (scientific) theory to another, according to Quine, is possible. It proceeds on the basis of rules or conventions that preserve the meaning of words, sentences, and ideas across cultural matrices. As Quine notes, "the parameters of truth stay conveniently fixed most of the time."[18] Like "the analytic hypotheses that constitute the parameter of translation," that is to say the possible interpretations of any sentence, the indeterminacy of translation does not remain fixed. Yet within a particular context, the ambiguities of any translation are diminished, if not removed momentarily; the use made of any translation concomitantly redefines and fixes its own parameters. Hence, the entire discussion of the indeterminacy of translation explicates those techniques—analytical hypotheses—for recognizing and fixing the parameters of indeterminacy.

Like the authority of the classical-canonical and modern-assessment views, the force and significance of linguistic analysis rests on its promise to provide a comprehensive vision of the world, even if the provision of comprehension is grounded in a multiplicity of methods. The contemporary-linguistic vision assumes the authority of resolving the most problematic aspects of knowledge acquisition and interpretation by referring to linguistic (cultural) conventions as if those conventions are configured prior to and independently of linguistic usage itself. Because the analysis of the incommensurability of theories is the analysis of discursive incommensurability of so-called scientific languages, it is connected directly to the problems of the indeterminacy of translation as defined by Quine.[19]

Where Quine uses a method of linguistic behaviorism, Kuhn relies on social psychology and sociology, and Feyerabend deploys what he calls the anthropological method. In their respective ways, each appeals to a specific point of reference from which their narratives are fabricated. These points of reference are presented in the name of *linguistic* analysis: natural or new starting points of analysis. Yet this analysis displays a metaphysical and ontological commitment to first principles masked by the ostensible reference to methodological and epistemological issues. Quine, Kuhn, and Feyerabend each claim to study the nature of science, or the structure of science, by examining the discourses of scientific theories—the different models, systems, and paradigms of science as they have been structured historically and linguistically. Just as Quine focuses on the parameters that fix truth and meaning in different languages and that enable translation from one language to another, so Kuhn and Feyerabend focus on the conventional and arbitrary decisions that factor into the comparison of theories and thus the translation of one scientific language into another.

By committing themselves to specific principles—conventional or pragmatic or anarchistic—that guide their methods and orientations, Quine, Kuhn, and Feyerabend present Meta-narratives. According to Kuhn and Feyerabend, the history of science can be understood as the *replacement of interpretations*, in which one interpretation is substituted for another, as opposed to viewing the history of science as the growth of scientific truths. However, the novelty of this recognition loses its force when Kuhn and Feyerabend, providing their own reconstruction of historical case studies, fail to recognize their sociological and anthropological reconstructions as replacements for classical, logical narratives. Hence, as self-declared replacements, the texts of Kuhn and Feyerabend perpetuate the struggle between Philosophical starting points through the incessant generation and replacement of interpretations.

For example, by distinguishing between "normal science" and the revolution characterizing a paradigm shift, Kuhn replaces the classical notion that scientific knowledge grows in a cumulative and uniform fashion with the notion that scientific progress proceeds by radical breaks between paradigms in the scientific framework. Standard histories (textbook histories) of science picture one generation of scientists building upon a foundation laid by the research of their predecessors. By contrast, because he accepts the notion of cumulative growth, Kuhn accounts for the anamolies that necessitate the radical break or (Gestalt) switch from one conceptual framework—psychological orientation—to one that resolves the anamolies.[20] Whence "scientific progress"!

Where Kuhn distinguishes between normal science and paradigm shifts, Feyerabend distinguishes between "new natural interpretations" and "familiar observational notions." At first, new natural interpretations appear "absurd," counterintuitive, producing "counterinductive assertions." As Feyerabend points out, Galileo was successful in replacing the Copernican model by using "propaganda" and "psychological tricks." Given their localized acceptance and retranslation into what were thought to be, undoubtedly, more appropriate idioms, the apparently unnatural interpretations and cosmological claims advanced by Galileo became familiar and quite natural. The consequence of this transformation or scientific "revolution" is the substitution of conceptual frameworks. Thus, Feyerabend's reconstruction of the move from Copernicus to Galileo is by its very character another replacement.[21]

Regardless of the differences that mark the criteria used in identifying the incommensurability of theories,[22] Kuhn and Feyerabend provide systematic accounts of how theories can or cannot be compared with one another—however anarchistic or conservative these accounts are supposed to be. For the purposes of this discussion, what is at issue is not the content of the specific rules or criteria used to determine incommensurability, but that such rules prefigure the unfolding of these narratives about incommensurability. In order to claim that a comparison is possible or impossible, certain elements that either contradict or correspond in content must be isolated by crossing the contextual boundaries of paradigms or theories. And this can be achieved only through an appeal to or application of specified criteria or first principles that are already enacted. The point of comparison, then, is to force a decision between conflicting interpretations or narratives — to rank order their explanatory predicative power.

The preference of one theory over another can be determined without effort if one theory is more comprehensive, presents a broader explanatory spectrum, or has greater predictive power than another. But incommensurability arises because theories cannot be compared according to such traditional criteria. Theories differ in ways that make a comparison, and eventually a choice, so difficult, if not impossible, that the only appeal available, according to Kuhn and Feyerabend, is to some set of conventions or arbitrary decisions, e.g., "aesthetic judgements, judgements of taste, metaphysical prejudices, religious desires, in short, *what remains are our subjective wishes* . . . "[23] Even though this conclusion appears to be a post-Philosophical declaration, its force remains that of a Philosophical principle.

On these grounds, a choice is never justifiable in "objective" — extratextual — terms. If the language of each theory or each narrative differs radically, with respect to the concepts and precepts that orient it, what would be the goal of such a comparison? Recognizing this problem and attempting to supply tentative solutions to it presumes a position of sovereignty, that is a position that remains outside the boundaries of comparison. In effect, it is a position from which the differences separating theories can be identified and comprehended. Quine, Kuhn, Feyerabend, and Rorty presume such a perspective; one positioned between discourses, between languages, where it can survey and assess the systematic or incessant generation of narratives.

The view of linguistic authority presented in the texts of Quine, Kuhn, and Feyerabend culminates in Rorty's commentary on the discursive conditions that permit exceeding the boundaries of the Philosophical culture. Each discourse constitutes a point or points of reference that indicate a way out of the labyrinth of narratives. As such, reference is made to an extralinguistic foundation, a foundation allegedly not part of the linguistic structure that orients the discourse. Moreover, it is a foundation that lies outside the range of its own critique. This is the perspective of the contemporary-linguistic view of authority. Deploying different sets of criteria, the contemporary-linguistic view is another disguise of the classical-canonical or modern-assessment views of authority. In effect, it turns on a transcendental appeal.

The Dissemination of Authority II: metanarratives

Up to this point in the discussion, we—GLO and RS in the position of narrators and storytellers as well as listeners, but not with the authority granted to authors or Philosophers—have generated metanarratives. These interpretations have isolated certain textual moments in order to articulate the assemblages—the convergence and divergence—of themes and problems traversing the histories of philosophy, science, technology, and literature. When any points of convergence and divergence (between science/technology, between philosophy/science, or between any texts or cultural matrices) have been fixed it has been for the purpose of moving to other points of concentration and difference. These points of reference do not arrest the incessant generation of narratives, because they are constituted, supplemented, and displaced by the dissemination of narratives.[24] Thus, no complete picture is presented—either from outside or inside the labyrinth or in between labyrinths. As is the case with *any* narrative, from our perspectives, the authority of ours is the authority of a metanarrative. That is to say, because all narratives are part of the labyrinths they constitute, the only authority narratives can declare is the authority of intervention and displacement—positioning themselves in between other narratives.

Further, our interpretations are not only *about* or *of* other texts, they are interpretations about *ourselves* or *themselves* as well, interpretations of the narratives we present. But the stories we tell about ourselves and the stories we tell about our stories are only some of the accounts that can be told either by us or by others. The accounts we have forged regarding the incessant generation of fictions, narratives, and labyrinths, as well as our accounts of displacement and dissemination, inevitably betray themselves. Our fictions are no more nor less authoritative than the others we have engaged. The strategies and techniques we have deployed in our treatment of other texts subvert the possibility of concentrated authority. In effect, we embrace the ambiguities, paradoxes, enigmas, and points of tension engaged throughout the generation of narratives and interpretations. And in doing so we announce and embrace the dissemination of authority—that is, the textuality of the cultural matrices we have articulated.

Where we have described the cultural matrices within which the interplay of domains takes shape, we also have talked about the labyrinths within which texts unfold. Each text is a switch point in the interplay of a multiplicity of domains: each text is intertextual already; its comprehension is possible only through the mediation of other texts. That is to say, aligning any text with certain scientific cultures is tantamount to aligning it with certain views of technology in accordance with certain linguistic orientations. It is for this reason that texts are relays within a series of relays; points of departure and never points of termination. Conceived as methodological, scientific, and theoretical, Bacon's texts, for example, express practical strategies for avoiding certain political and linguistic follies. In a similar fashion, Foucault's texts are commonly treated as presenting an etymological and textual archaeology or genealogy of cultural systems, but they are, as well, inextricably bound to the theoretical debates that dominate the history and methodology of science and technology. Moreover, though identified by the emphasis placed on a pragmatic approach to social and educational practices, Dewey's texts speculate on the technological/theoretical means by which culture is understood and can be reconstructed in the future.

In our readings of various texts, we have insisted that texts, as well as the themes and problems they articulate, are part and parcel of an incessant generation of fictions and narratives. For us, each text—including our own—is a metanarrative, an account of the interplay of texts, the reciprocity of interpretive interventions, and the generation of interpretations of interpretations.

Metanarratives enable different perspectives that traverse other narratives and their histories within a multiplicity of interpretations. Because there is no foundation outside the generation of narratives to authorize a metanarrative, there is no primacy of place at the beginning or the end. There is no first nor final word; there are only words, always connected to and presupposing other words. There is only the incessant generation of narratives. All narratives, then, are platforms, but platforms always staged in intermediary roles.[25] Mediation of this sort does not function on the possibility of overcoming or superseding any other narratives. Even though mediation segments the assemblage of cultural domains, it constitutes the conditions of reciprocity. Just as mediations make possible textual differentiation, so textual identifications are the possibilities of mediation.

We present thematically mediated metanarratives. There is certainly a multiplicity of themes that could have been used, and certainly there are other metanarratives that could have been forged. But the one theme to which our narratives return constantly is this: linguistic performance is the only warrant of authority. (It is this theme that makes it possible to apprehend, albeit in a provisional manner among other comprehensive techniques, other issues and questions that might be thought otherwise incommensurable.) In other words, there is no authority outside the performative: there is no authority outside discursive declarations. In effect, a text can appeal only to the practices and techniques used in its fabrication as the basis of its rule. Conceived in this way, language does not constitute a medium for the presentation of some object that lies outside its apparatus. Instead, its usage creates the object, as well as the frameworks within which the object of any discourse is presented.

To say "use creates" is to privilege a pragmatic problem or orientation. But like any discursive privileging it is without foundation: it is ad hoc and post hoc. Thus, we are convinced that only the effects produced and mediated by any narrative account, and which come to mediate the production of textual effects, characterize the alleged nature of any narrative account. For example, a text is interpreted as a "scientific" treatise, "Philosophical" investigation, or a "personal confession," as Nietzsche would say, regardless of the text's techniques, its declarations, and the principles that structure it. Likewise, an African mask is interpreted as an "anthropological artifact" worthy of scientific investigation, at one point, and as a work of "art" appropriate for museum exhibition, at another point. To say "use creates" is to renounce any notion of an "essence," in the spirit of Wittgenstein's philosophical declarations.[26]

We have noted already that counterfeiting underlies all textual declaration and presentations.[27] Texts represent nature or the world vis-à-vis metanarratives. It is the *meta* character of these narratives, that is the insertion in between representations, that orders counterfeiting. Moreover, it is the techniques of counterfeiting according to which science, technology, and philosophy are ordered, institutionalized, and thus canonized. "Use creates," then, is one of the rules legitimating our discursive counterfeits. Like other specific rules articulated for the purpose of generating or disseminating systematic fictions, the rules we use are ad hoc and post hoc. No rule has warrant prior to its use as a legitimating device for any declaration (about the world, about science/technology, about rules). And because a rule is nothing more nor less than a heuristic device, its warrant is fleeting, transformed, and supplemented with each deployment.

Since rules legitimate themselves only through use, there can be no appeal to some concentrated, transcendental or transcendent authority. In other words, to appeal to any rule is to use the rule, that is to forge the rule in accordance with the dissemination of authority. When rules are fixed they are not fixed permanently nor are they fixed outside a specific pragmatic textual context. A great deal of counterfeiting is required in order to anchor rules or to fashion them in an "ideal" form. This situation interests us because of its relevance to current pedagogical practices. Readers—students and teachers—are encouraged to fix rules, to change the principles, or invent different ones as the context demands, that guide their action, thought, and discourse. In effect, they are encouraged to recognize the fabrication of rules as an inevitable effect of reading/writing—the fabrication of narratives, interpretations, and fictions. Furthermore, they are encouraged to recall how their interpretations resituate themselves within the labyrinths—scientific/technological, theoretical/practical, etymological/textual—that orient the exchanges between narratives, thought and action, and concept and practice.

Anámnēsis, Interpretations, and Performances

We have engaged in and assembled a variety of exchanges. Any set of relations that regulate exchanges constitutes an economy. Our engagements and exchanges with other narratives have established the rule "use creates," among others, in order to comprehend at least one economy among a multiplicity of economies and labyrinths. The effects of these exchanges have been discursive displacements and the dissemination of authorities. Displacement and dissemination are inextricably intertwined with the rule instituted within labyrinths. With the replacement of one narrative with another, there remains a concentration of authority and meaning. Moreover, there remains an illusion of a center, that point of concentration that is presented as a grand narrative. But with discursive displacement, there remains nothing of a center—only segments, switch points, and platforms, or what Derrida and Lyotard call "spacings."[28]

There remain only fragments, traces, or relays alluding to other connections. In effect, there is only the production of narratives, fictions, representations as connecting traces in the generation of other accounts. So, when we have alluded to a metanarrative, we envision a comprehensive technique for the possible regulation of an economy, in the recognition that a single economy can be isolated amidst many but never insulated from the general effects of other economies. Just as our rules are privileged only in a nominal sense, so the economies with and within which we operate are privileged only in a fleeting fashion.

We have identified and engaged certain techniques involved in textual displacement. We have tried to show that these techniques are employed, in various ways, in certain texts privileged in the histories of philosophy, science, technology, and literature. As far as we are con-

cerned here, these techniques are interpretive techniques. Every interpretation is a point of intersection. That is to say, every interpretation is
a point at which connections are made, and underlying the possibilities of these connections is a form of recollection. But it is not the kind
of recollection that returns to the past, to the memory of history in order to resuscitate the canons of a particular enlightenment.

Recollection, understood broadly, is and has been used commonly
as a technique for substantiating certain claims or giving shape to certain declarations, as a practice for connecting certain selected fragments found in other seemingly unrelated texts. As such, the practice
of recollection constitutes "recollection" (anámnēsis) as a rule, the use
of which is regulated by the connections made within a specific context. In terms of the effects produced, recollection is another name for,
another form of, interpretation.

As a rule, recollection is a performance that situates itself as an interpretation on a particular stage.[29] To this extent, like other interpretations and performances, recollections present the stage on which
their authority is and will be declared. Defining the contextual parameters of a performance, one stage differs from other stages. Its specific
use—the performance of an interpretation—is the difference. And
since the stage is the site of recollection, it resists the attempt of any
performance to cast itself as an "original." At best, the words recited
are words picked up here and there, for a particular purpose, and re-
phrased and paraphrased in this or that manner, within a specific context. Like etymologies, recollections present fictional discourses—or
as the pragmatics of a situation would demand, opportunities—within
which legacies counterfeit a tradition and, thus, remake the labyrinths.
Under such conditions, every word, every fragment of every text,
lends itself to the undecidability of interpretations.

The incessant generation of interpretations (narratives, fictions,
etc.) signifies the dissemination of authority and the displacement of
(con)textual boundaries. By focusing on the performative aspect of interpretation, it is clear to us that the issue is not who speaks, but that
something is declared and requires a response that, in turn, resituates
what has been said. In this respect, language is not the language of
representation that characterizes the Meta-narratives of the tradition.
Instead, language is only presentation, the provision of connections in
the remaking of possible traditions.

The cultural matrices we have identified and with which we are concerned can be traced only as points of departure. Science/technology,
for example, does not present a fixed matrix of interpretations or a definite field within the limits of which certain interpretations are declared

legitimate. Even when rules are articulated within the bounds of any specific matrix, or in between matrices, they are certain effects produced in and through interpretive interventions or performances. They are not the kind of criteria that will legitimate transcending and replacing one interpretation or Philosophical culture with another, as is the case with Rorty. Instead, rules, as we have indicated, are in constant transformation, always changing with the demands of a given context. Narratives or interpretations always seek the rules that will legitimate their declarations and the effects they might produce. Their articulation stages the announcement of different rules as well as disseminates the authority claimed in the declaration itself.

To be sure, the incessant generation of interpretations, narratives, fictions, systems, and so on presupposes and embraces the concept of plurality. However, it is not the kind of plurality or pluralism championed by Rorty, for example, nor the criteria-bound interpretive pluralism promoted by certain adherents of the social study or sociology of science, nor the hierarchical pluralism advocated in certain trends of contemporary literary criticism. When Rorty distinguishes the post-Philosophical and Philosophical cultures he announces his preference for the post-Philosophical *over* the Philosophical. Notwithstanding his desire for philosophical pluralism, Rorty's *preference* for the post-Philosophical establishes a rank ordering within the plurality. Rorty uses the post-Philosophical as a comprehensive Meta-narrative, thereby *eliminating* the plurality of cultures.[30]

Today, certain sociologists and philosophers study the scientific enterprise by examining and stressing the social context of that enterprise (as exemplified by the so-called Edinburgh School) and, by doing so, introduce the concept of interpretive pluralism, with respect to both texts and actions, into the picture. Rejecting the claim that the scientific enterprise can be judged only with reference to its particular "methodology," Steve Woolgar and Steven Yearly, for example, emphasize the social environment in which research is constructed and attempt to distill a plurality of interpretations based on the "facts" supplied by textual records and laboratory activities. In distinguishing between kinds of interpretations, where interpretation is a "representational" device, they declare a preference for a sociological or anthropological approach that will reflect the "real" nature of science. Arguing for the comprehensive character of "enthnographical" or "constructionist" accounts, they privilege their accounts on the basis, as Lyotard would say, of what they do, while at the same time, they remain devoted to "the very idea" of "science."[31]

Within contemporary literary criticism (as practiced by the so-called Chicago School), pluralism holds a privileged position among other standard approaches to the question of textual interpretation. The pluralism advocated includes dogmatism or any other approach that is construed to be more narrow than the proliferation of interpretations. But by subsuming dogmatism within the framework of comprehensive *critical* pluralism, dogmatism is denied an independent position — there is only pluralism — and pluralism is not seen as one of many other interpretive frameworks. Pluralism legitimates itself by censure.[32]

Whether one talks about the philosophical pluralism of Rorty's pragmatism, or the ethnographical pluralism of recent developments in the sociology of knowledge, or the critical pluralism of contemporary literary criticism, the pluralism does not entail a dissemination. In effect, these interpretive practices that all at once fall under the heading of *pluralism* gravitate toward the centralization of authority. In other words, the discourse of pluralism produces another principle of unification, where a foundation is established, that is a Meta-narrative, and where differences are overshadowed. Thus, the appeal to interpretive pluralism is yet another disguise for an appeal to hierarchies, standards, and criteria. And, as we have indicated, any appeal to something outside the production of interpretive accounts is an appeal to authority consonant with the classical-canonical model.

There are several ways to legitimate our own "preference" for the discursive displacements of boundaries and the dissemination of authorities — textual or other. First, we are convinced that the production of metanarratives is all there is. That is to say, there is no central nor external point of reference, no position of sovereignty outside the labyrinths of narratives that would establish such a point. In effect, there is no re-presentation of representation (*Vorstellung*).

Second, we are convinced that the only sense that we can make of words and concepts such as *meaning, truth, reality,* and *self* arises from the *use* made of them within the labyrinths of narratives. In this regard, the languages of the labyrinths, that is to say the discourses used to construct, portray, and explore the labyrinths, are not re-presentational. Instead, they are declarative and performative: they are idiomatic or presentative (*Darstellung*).

Third, we are convinced that in spite of, and because of, the dissemination of authority, our self-legitimating claims and accounts can be assigned no more nor less authority than the classic-canonical, modern-assessment, or the contemporary-linguistic narratives. The stake we have in the dissemination of authority is neither fixed nor final. We understand that by stating a preference we are not committed exclusively to the labyrinthine accounts we have provided. Nor does it commit us to a pledge of abstinence vis-à-vis other interpretive strategies and techniques. We acknowledge that the differences in context and purpose will require different metanarratives. Indeed, we understand that there are times when assuming a posture resembling the canonical view of authority is entirely appropriate as a technique of inquiry or as a mode of/for presentation. Any choice of interpretive techniques and strategies is, practically, a choice between shorthand techniques—specific means of expression (énoncé) designed for purposes demanded within a particular context.

And finally, because we do not presume to present a Meta–narrative, we do not know if our own accounts and interpretations transcend or exceed other meta–narratives involved in the self-structuring practices of labyrinths. Besides, to re–call an earlier declaration, the issue remains not *who* speaks but that something has been announced in response to something else—and that response solicits acknowledgment.

Pragmatic Maxims, Performances, and Pedagogy

In *Ion* Socrates engages Ion, a performer of Homeric recitals, in a dialogue regarding certain interpretative practices—techniques and strategies—deployed in the recitation of the stories about the gods as they are narrated by the poet Homer. Ion agrees with Socrates: he is nothing more nor less than an interpreter of other interpreters, i.e., the poets and Muses, just as Socrates is himself such an interpreter (535 a). What interests Socrates, then, are the different techniques employed by interpreters in the name of the "original" interpreters for the purposes of producing certain effects under the sundry conditions and circumstances of their performances (535 d-e). As Ion acknowledges:

> As I look down at them from the stage above, I see them, every time, weeping, casting terrible glances, stricken with amazement at the deeds recounted. In fact, I have to give them very close attention, for if I set them weeping, I myself shall laugh when I get my money, but if they laugh, it is I who has to weep at losing it (535 e).

Through Ion, Plato's Socrates shows how Homer's narratives solicit different responses, different performances and uses, regardless of what might be identified as the original intent of the poet or what the interpreter claims the poet meant to say by way of divine inspiration (530 c). Moreover, Socrates shows how the interpretive strategies used become a multiplicity of means for staging a story about a particular issue.

Given their context, the appropriateness of Ion's theatrical performances is linked to his qualifications to interpret Homer. By his own
admission, his qualifications—his authority—rest on the declarations
of a group of "qualified" judges who were convinced by Ion's performance that he was the most "conversant with many excellent poets,
and especially with Homer, the best and most divine of all" (530 b).
Moreover, the judges' decision is consonant with Ion's assessment of
his performance: "In my opinion I deserve to be crowned with a
wreath of gold by the Homeridae" (530 e). However, no interpretation—of Homer or Plato or Socrates—is bound by the stage on which it
is performed. Interpretations are always interpretations of interpretations. As such, they supply intermediary cases for exploiting the values
and conditions of the culture in which they are situated. Hence, there
are nothing but intermediary performances always staged in between
one another and thereby always situating one another.

To be in between narratives, then, is to be at the borders of narratives.
It is always to be in the positions of looking for ways or techniques to create another assemblage that will comprise various segments re-collected
from selected narratives. To this extent, re-collection is not a technique of
representation. Indeed, re-collection is the form of thought that "corresponds" to the "reality" of discursive displacement and dissemination. It
is, then, the re-collection of other recollections; the transfiguration of
legacies and enlightenments as they are reiterated in different contexts.

Furthermore, to be *in between* narratives is to be fastened to and,
simultaneously, detached from narratives that have been articulated as
well as those that remain fragmented. Discursive borders are never
fixed nor static; to be in between is to be situated at many borders all
at once. It is to be engaged in the reciprocity that conditions the incessant generation of narratives. We do not work with only one idiom (or
thematic assemblage of a tradition) at a time: but we do not know how
many idioms (thematic assemblages) are at work at any one time, nor
exactly how they will be used or interpreted.

So, in order to be *outside* a particular tradition or legacy, one must
remain *inside*. As one will recall, it is this situation that Socrates describes with the allegory of the cave (*Republic* 514-17). One must use,
that is to say *reinscribe*, the idioms and concepts of a particular tradition for a purpose other than what the canons of the tradition declare.
Determining whether one is in or out is a function of the use made of
the images, beliefs, and theories appropriated from the tradition.
Thus, in order to be inside a particular tradition or "cave" one must
recognize it as a *return*, where the borders are used to designate an
outside or those conditions that necessitate recollecting the traditional

idioms in different ways. In effect, by the necessity of use one would be both but neither in/outside any tradition or legacy.

To be sure, there are explicit Socratic themes that inform our understanding of how the borders or limits of narratives are incessantly transformed or transfigured. And to be sure, these themes are pedagogical in orientation. Like Socrates/Plato, we recognize, and this is one hypothesis we celebrate, that "seeking [inquiry] and learning are in fact nothing but recollection" (*Meno* 81 d). In similar fashion, we recognize that "there is no such thing as teaching, only recollection" (82 a). One consequence of these hypotheses, one that presents re-collection as a form of experimentation, is the following declaration: "if there are no teachers, neither are there disciples [students or learners]" (96 c). Thus, we do not profess to be teachers nor disciples of particular canons or sacred texts. We only profess to enjoy the experimentation of re-collection.

Our experimentations have led us to *re-call* that the re-cognition of any problem as a problem, any question as a question, and any issue as an issue is a textual/pragmatic occasion. First, we have emphasized, again and again, that the determination of nature, reality, thought, society, and self is discursive. Making reference to any of these fundamental concepts always involves making reference to some other textual determination. There is nothing outside textual accounts.

Second, we recognize that not only are there only textual accounts, there is a multiplicity of these accounts. This multiplicity is a consequence of counterfeiting, that is to say the incessant generation of narratives—fictions, visions, representations, stories, discourses, models, systems, and texts. These accounts can be—have been and will be—construed to satisfy the demands of particular cultural needs at some undetermined moment. As such, they can be understood and presented in a unified, totalized manner as if to present a tradition or privileged idiom.[33]

Third, we recognize that the incessant generation of narratives is the production of labyrinths, but where a labyrinth is nothing other than the assemblage of narratives involved in the production of labyrinths. With the multiplication of textual accounts—the generation of idioms, linguistic linkages endemic to a specific culture—there is also a multiplication of labyrinths. Isolating a labyrinth, or one of its economies, is analogous to giving primacy of place to a tradition or idiom or narrative. That is to say, only specific use of another narrative warrants its own declaration of authority as providing the framework for comprehending all other accounts.

Fourth, we recognize re-collection as an instance of textual dissemination or discursive displacement. Dissemination and displacement are functions of the language of narratives that document the primacy of their own modes of presentation, such as books, laboratory experiments and reports, pamphlets, novels.[34] But the languages of architecture, film, music, painting, and the theater also involve—produce and are products of—the practices of dissemination and discursive displacement. Just as the language of philosophy has been used to comprehend the language of science/technology, we use the language of art history or aesthetics, for example, to comprehend the language of a musical score or a painting, as if the one could faithfully re-present the other. But such an imposition of linguistic frameworks or cultural matrices is itself a demonstration of the displacement and betrayal of traditional domains. As a consequence, there are no extratextual points of reference, nor centers of authority. A *text*—a book, an artifact, a painting, a film, or any object of cultural analysis—presents or announces itself and others as relay(s) undergoing constant transformation and departure.

Fifth, we recognize that every text is always in between other texts. Bordering other texts is to be an intermediary text, a text that situates itself in terms of how it re-situates other texts. With every interpretation there is a realignment of passageways within the labyrinths. As a consequence, a fixed discourse, a frozen tradition, or a labyrinth with a given foundation becomes one among many other possibilities. In other words, like K. in Kafka's *The Castle* or *The Trial*, we find ourselves always at an intersection—always involved in the proliferation of borders and passageways, realizing that every door swings in/out, exposing/suspending an assemblage of borders. And yet we always attempt to find an escape, a way out.[35]

And sixth, we recognize there is no re-cognition without experimentation. Re-cognition is brought about only through re-collection: the appropriation and use of certain textual themes and techniques according to the context in and purpose for which they are to be used. We realize that our experiments—whether they are classified as analyses, critiques, or interpretations—have unfolded in this fashion. And as a result, we realize that the narratives we have recalled, and the themes and techniques that we have engaged in specific ways, are themselves recollections or the occasions for recognition and experimentation. In this respect, an experiment neither guarantees nor warrants any particular consequences. If anything, as with any experiment, the effects of a narrative cannot be predetermined nor legitimated by appeal to the intention or declarations of an "author."

At best, education can only be undertaken as an experiment—an attempt to engage others by recollecting—responding or reacting—to the constellation of textual phrasings or connections. The perpetuation of a specific tradition, or fostering disciples, is not the issue. The issue of education, in this context, is re-collection: recalling the uses made of certain interpretive techniques and strategies, the purposes for which these techniques have been employed, and the recognition of the differences these techniques and strategies have produced. The question of education, then, is not one of what *should* be learned or taught, or what an author *means* or *intends*, but rather what *effects* a particular narrative generates.

If education is practiced according to the former position, then there is a sense in which education becomes a technique for domination or liberation, a form of dogmatism or enlightenment. There is a sense of oppression *and* liberation that accompanies being taught what one should know as proscribed by a privileged method. On the other hand, if education is practiced as experimentation, no appeal can be made to what should be known nor how it should be known. Without the security of a foundation, or an identifiable authority associated with a master narrative, one can experience a sense of anxiety and frustration. But, as Nietzsche would declare, the anxiety and frustration of an experiment (*Versuch*) are to be celebrated and feasted upon.[36] And yet, like Nietzsche, we recognize that in spite of its so-called openness and freedom, experimentation can betray its own techniques by returning itself to the abyss of formal education. Thus, like Nietzsche, we also recognize how difficult it is to celebrate.

Is there nothing "new" to learn? Does our "knowledge"—philosophical, scientific, technological, artistic, . . . —remain the "same"? Given the Socratic/Nietzschean orientation of these comments, the answer is a resounding no! In *Ecce Homo*, Nietzsche claims that "Ultimately, nobody can get more out of things, including books, than he already knows. For what one lacks access to from experience one will have no ear."[37] So, how does one learn something "new," hear or see something "different"? One creates it. But how is this fabrication possible? What legitimates its presentation? Its presents? The experimentation of re-collection. Recollection is seeing, hearing, interpreting, . . . something new or different by exposing particular canonical interpretive strategies and techniques in another context, for another set of purposes. Exposition, then, is its own legitimation. It incorporates both the acceptance and betrayal, that is to say, the learning and unlearning, of traditions. Every exposition is a form of relearning. The tools of the traditions are reconstructed and reused continuously, so that the presentation of the "old" is always the presentation of the "new," and the new represents the old.

The Last Words: Reading First Lines

"The exploration of what is called 'science, technology, and society' assumes an interlacing of the multiple cultural and linguistic dimensions of contemporary life. . . . "

"Two interrelated claims have been advanced regarding the interplay of science and technology. . . . "

"Negotiating the intricate assembly of any labyrinth (or texts) requires dismantling and abandoning, or at the very least suspending, certain classical categories and binary oppositions. . . . "

"Thus far, the texts and themes examined have been used to achieve a particular focus: to explore the roles language plays in determining and situating science and technology within particular cultural matrices. . . . "

"All narratives, whether they are cast as 'philosophical' inquiries, 'scientific' treatises, or 'literary' texts, are inextricably bound to a labyrinth of interpretations, that is a web of other fictions, other narratives. . . . "

Ours is a book both about and not about *science and technology*. But what has been learned, if anything, from our experiment? We have learned that science and technology, like any textual assemblage, is apprehended and comprehended only through its use. And use orders a different concentration of significance or value according to the cultural topography into and out of which it is inscribed. Moreover, we have learned that if our discourse has as its subject the culture of science and technology, it has also as its subjects the cultures of *history, philosophy, politics, economics, literature,* and *education.*

Notes

1. Rorty, *Consequences of Pragmatism*, xxxviff.
2. Ibid., xxxvi.
3. Ibid., xxxvii.
4. Ibid., xli.
5. Richard Rorty, *Philosophy and the Mirror of Nature* (Princeton: Princeton University Press, 1979), 367ff.
6. Rorty, *Consequences of Pragmatism*, xli. As Rorty notes, "On the pragmatist account, a criterion (what follows from the axioms, what the needle points to, what the statute says) *is* a criterion because some particular social practice needs to block the road of inquiry, hold the regress of interpretations, in order to get something done." In effect, by issuing a pragmatic criterion Rorty attempts to arrest "the regress of interpretations," or avoid an accusation of relativism, in order to "get something done," i.e., to find a point of agreement for the generation of commentaries. See also xliii.
7. Ibid., xxxvi.
8. Rorty, *Philosophy and the Mirror of Nature*, 316.
9. Compare Ibid., 316 n. 1.
10. This question is based on the discussion that appears in chap. 4.
11. Willard Van Orman Quine, *Word and Object* (Cambridge: MIT Press, 1960), 28.

12. Compare Otto Neurath, "Protocol Sentences," *Logical Positivism*, ed. A.J. Ayer (Boston: Free Press, 1959), 199-208; and Rudolph Carnap, *The Logical Syntax of Language* (Paterson, N.J.: Littlefield, Adams, and Co., 1959), 222-33.

13. Compare Quine, *Word and Object*, 79. Quine writes: "The indeterminacy of translation has been less generally appreciated than its somewhat protean domestic analogue. In mentalistic philosophy there is the familiar predicament of private worlds. In speculative neurology there is the circumstance that different neural hookups can account for identical verbal behavior. In language learning there is the multiplicity of individual histories capable of issuing in identical verbal behavior. Still one is ready to say of the domestic situation in all postivistic reasonableness that if two speakers match in all dispositions to verbal behavior there is no sense in imagining semantic differences between them. It is ironic that the interlinguistic case is less noticed, for it is just here that the semantic indeterminacy makes clear empirical sense."

14. Ibid., 28.

15. Ibid., 27.

16. Ibid., 34.

17. Ibid., 75-76.

18. Ibid., 76.

19. Compare Kuhn, *The Structure of Scientific Revolutions*, 202ff.; and Feyerabend, *Against Method*, 287.

20. Kuhn, *The Structure of Scientific Revolutions*, 66-135.

21. Feyerabend, *Against Method*, 47-108.

22. Compare Paul Feyerabend, "Changing Patterns of Reconstruction," *British Journal of the Philosophy of Science* 28 (1977): 363-67.

23. Feyerabend, *Against Method*, 285.

24. The notion of "dissemination" in use throughout our discussion is spurred by the writings of Jacques Derrida. Compare Jacques Derrida, *Of Grammatology*, trans. Gayatri Chakravorty Spivak (Baltimore: Johns Hopkins University Press, 1975), see especially 269-316; and Jacques Derrida, *Dissemination*, trans. Barbara Johnson (Chicago: University of Chicago Press, 1981), see especially, 3-59 and 289-366.

25. See Wittgenstein, *Philosophical Investigations*, pt. 1, sec. 122. "A main source of our failure to understand is that we do not *command a clear view* of the use of our words. — our grammar is lacking in this sort of perspicuity. A perspicuous representation produces just that understanding which consists in 'seeing connexions.' Hence the importance of finding and inventing *intermediate cases*."

26. Ibid., pt. 1, sec. 65, 92ff, and 532.

27. See chap. 2, "Fictional Visions of Science and Technology."

28. Compare Derrida, *Dissemination*, 208ff, and *Of Grammatology*, 203ff and 232ff. Derrida writes:

> What cannot be thus represented by a line is the turn (trick/trope) of the return when it has the bearing of re-presentation. What one cannot represent is the relationship of representation to so-called originary presence. The representation is also a de-presentation. It is tied to the work of spacing. (*Of Grammatology*, 203)

And:

> Language could have emerged only out of dispersion . . . .This dispersion should no doubt be overcome by language but, for that very reason, it determines the *natural condition* of language.

The natural condition: it is remarkable that the original dispersion out of which language began continues to mark its milieu and essence. That language must traverse space, be obliged to be spaced, is not an accidental trait but the mark of its origin. In truth, dispersion will never be a past, a prelinguistic situation in which language would certainly have been born only to break with it. The original dispersion leaves its mark within language. We shall have to verify it: articulation, which seemingly introduces difference as an institution, has for ground and space the dispersion that is natural: space itself. (*Ibid.*, 232)

Also, see Jean-François Lyotard, *Peregrinations: Law, Form, Event* (New York: Columbia University Press, 1988), 31.

It's [the blank, the space] the emptiness, the nothingness in which the universe presented by a phrase is exposed and which explodes at the moment the phrase occurs and then disappears with it. The gap separating one phrase from another is the "condition" of both presentation and occurences, but such a "condition" remains ungraspable in itself except by a new phrase, which in its turn presupposes the first phrase. This is something like the condition of Being, as it is always escaping determination and arriving both too soon and too late.

29. In using the term *performative*, we are appropriating a term defined by J.L. Austin, *How to do Things with Words* [1962] (Cambridge: Harvard University Press, 1975). Austin distinguishes the performative as one kind of speech act from the constitutive as another kind. The distinction is one between "doing and saying" (47; 8-11). We presuppose this distinction, but, like Derrida, we take license in demonstrating that even "saying," that is statements that are true or false, is a kind of "doing," a declaration that produces undetermined effects. Compare Derrida, "Signature Event Context," *Glyph* 1: 172-97; and "Limited Inc a b c . . . ," *Glyph* 2, (1977): 162-254.

30. Compare Rorty, *Consequences of Pragmatism*, xlff.

31. Compare Steve Woolgar, *Science: The Very Idea* (London and New York: Tavistock Publications, 1988), 11-14 and 83-96. Woolgar's primer draws from the following works: Barry Barnes, *Scientific Knowledge and Sociological Theory* (London: Routledge and Kegan Paul, 1974); David Bloor, *Knowledge and Social Imagery* (London: Routledge and Kegan Paul, 1976); H. M. Collins, *Changing Order: Replication and Induction in Scientific Practice* (London: Sage, 1985); and Bruno Latour and Steve Woolgar, *Laboratory Life: The Construction of Social Facts* (Princeton: Princeton University Press, 1986). Compare also Steven Yearley, *Science, Technology, and Social Change* (London: Unwin Hyman, 1988), see especially 16-66 and 181-86. For example, Yearley claims that "The constructionist clearly accepts that science and technology are the best resources we have for dealing with the natural world" (186).

32. Compare W. J. T. Mitchell, "Pluralism as Dogmatism," *Critical Inquiry* 12 (Spring 1986): 494-502. This issue of *Critical Inquiry* is devoted to the topic "Pluralism and its Discontents."

33. See "The Practical Context of Current Issues," in chap. 1.

34. Compare Joan Digby and Bob Brier, ed., *Permutations: Readings in Science and Literature* (New York: Quill, 1985). This collection of readings joins the works of novelists and scientists as they comment on similar topics, such as science, chemistry, biology, etc.

35. Compare Gilles Deleuze and Félix Guattari, *Kafka: Toward a Minor Literature*, trans. Dana Polan (Minneapolis: University of Minnesota Press, 1986). The proliferation of hallways and doors is a theme given extensive treatment in this analysis of Kafka.

36. Compare Nietzsche, *The Birth of Tragedy*, especially the "Preface to Richard Wagner" and the preface of 1886, "Attempt at Self-Criticism."

37. Friedrich Nietzsche, *Ecce Homo* [1888], trans. Walter Kaufmann (New York: Vintage Books, 1969), 261. Nietzsche continues: "This is, in the end, my average experience and, if you will, the originality of my experience. Whoever thought he had understood something of me, had made up something out of me after his own image—not uncommonly an antithesis to me; for example, an "idealist"—and whoever had understood nothing of me, denied that I need be considered at all."

SELECTED BIBLIOGRAPHY

SELECTED BIBLIOGRAPHY

Agassi, Joseph. *Technology: Philosophical and Social Aspects*. Dordrecht: Reidel, 1985.

Austin, J. L. *How to do Things with Words* [1962]. Cambridge: Harvard University Press, 1975.

———. "Performative Utterances." In *Philosophical Papers*, 2d ed., eds. J. O. Urmson and G. J. Warnock. Oxford: Clarendon Press, 1970.

Ayres, Clarence E. *Science: The False Messiah* [1927] Clifton, N.J.: Augustus M. Kelley, 1973.

Bacon, Francis. *The New Organon* [1620], ed. Fulton H. Anderson. New York: Macmillan, 1985.

Barnes, Barry. *Scientific Knowledge and Sociological Theory*. London: Routledge and Kegan Paul, 1974.

Bernstein, Richard J. *Beyond Objectivism: Science, Hermeneutics, and Praxis*. Philadelphia: University of Pennsylvania Press, 1983.

Bloor, David. *Knowledge and Social Imagery*. London: Routledge and Kegan Paul, 1976.

Borgmann, Albert. *Technology and the Character of Contemporary Life: A Philosophical Inquiry*. Chicago: University of Chicago Press, 1984.

Borgmann, Albert, and Carl Mitcham. "The Question of Heidegger and Technology: A Critical Review of the Literature," *Philosophy Today*, Summer 1987 (Special Issue): 98-191.

Boyle, Godfrey, David Elliot, and Robin Roy. *The Politics of Technology*. London: Open University Press, 1977.

Carnap, Rudolph. *The Logical Syntax of Language*. Patterson, N.J.: Littlefield, Adams, and Co., 1959.

Clarke, Arthur C. *July 20th, 2019: Life in the 21st Century*. New York: Macmillan, 1986.

———. *Profiles of the Future*. New York: Dell, 1968.

———. *2001: A Space Odyssey*. New York: Dell, 1969.

Cohen, I. B. *Revolution in Science*. Cambridge: Harvard University Press, 1985.

Collins, H. M. *Changing Order: Replication and Induction in Scientific Practice*. London: Sage, 1985.

Comte, Auguste. *The Second System: Système de Politique Positive* [1851-54]. In *Auguste*

Comte and Positivism: The Essential Writings, ed. Gertrude Lenzer. New York: Harper & Row, 1975.

Defoe, Daniel. *Robinson Crusoe*. New York: Bantam Books, 1981.

Deleuze, Gilles. *Cinema I: L'Image-Mouvement*. Paris: Les Éditions de Minuit, 1983.

———. *Cinema I: The Movement-Image*, trans. Hugh Tomlinson and Barbara Habberjam, Minneapolis: University of Minnesota Press, 1986.

Deleuze, Gilles, and Félix Guattari. *Kafka: Toward a Minor Literature*, trans. Dana Polan. Minneapolis: University of Minnesota Press, 1986.

Derrida, Jacques. "Déclarations d'Independance." *Otobiographies: L'enseignement de Nietzsche et la politique du nom proper*. Paris: Éditions Galilée, 1984.

———. *Dissemination*, trans. Barbara Johnson. Chicago: University of Chicago Press, 1981.

——— "Limited Inc a b c . . . ," *Glyph* 2, (1977): 162-254.

———. *Margins of Philosophy*, trans. Barbara Johnson. Chicago: University of Chicago Press, 1981.

———. *Of Grammatology*, trans. Gayatri Chakravorty Spivak. Baltimore: Johns Hopkins University Press, 1975.

———. "Signature Event Context," *Glyph* 1, (1977): 172-97.

Descartes, René. *The Discourse on Method of Rightly Conducting the Reason and Seeking for Truth in the Sciences*. Vol. 1 of *The Philosophical Works of Descartes*, trans. Elizabeth S. Haldane and G. T. R. Ross. Cambridge: Cambridge University Press, 1981.

Dewey, John. *Experience and Nature* [1929]. New York: Dover, 1958.

———. *Reconstruction in Philosophy* [1920]. Boston: Beacon Press, 1948.

———. *The Quest for Certainty: A Study of the Relation of Knowledge and Action* [1929]. New York: Capricorn Books, 1960.

Dickens, Charles. *Hard Times*. New York: Bantam Books, 1981.

Digby, Joan, and Bob Brier, eds. *Permutations: Readings in Science and Literature*. New York: Quill, 1985.

Douglas, Mary, and Aaron Wildavsky. *Risk and Culture: An Essay on the Selection of Technical and Environmental Dangers*. Berkeley: University of California Press, 1982.

Duhem, Pierre. *The Aim and Structure of Physical Theory* [1906], trans. Philip P. Wiener. Princeton: Princeton University Press, 1954.

———. *To Save the Phenomena: An Essay on the Idea of Physical Theory from Plato to Galileo* [1908], trans. Edmund Dolan and Chaninah Maschler. Chicago: University of Chicago Press, 1969.

Durbin, Paul (1978-88) and Frederick Ferré (1988-), eds. *Research in Philosophy and Technology*. Westport, Conn.: JAI Press.

Durbin, Paul, ed. *Philosophy of Technology*. Dordrecht: Reidel, 1987-.

Encyclopedia Britannica, 11th ed., vol. 26, New York: Macmillan, 1911.

Ellul, Jacques. *The Technological Society*, trans. John Wilkinson. New York: Alfred J. Knopf, 1964.

Ferré, Frederick. *Philosophy of Technology*. Englewood Cliffs: Prentice Hall, 1988.

Feyerabend, Paul. *Against Method: Outline of an Anarchistic Theory of Knowledge*. London: NLB/Verso, 1975/78.

———. "Changing Patterns of Reconstruction," *British Journal of the Philosophy of Science* 28 (1977): 363-367.

Foucault, Michel. *The History of Sexuality, Vol. I: An Introduction*, trans. Robert Hurley. New York: Vintage Books, 1980.

———. *The Order of Things: An Archeology of the Human Sciences* [1966]. New York: Vintage Books, 1973.

_____ . *This is Not a Pipe*, trans. and ed. James Harkness. Berkeley: University of California Press, 1983.

Fuller, R. Buckminster. *Critical Path*. New York: St. Martin's, 1981.

_____ . *Earth, Inc.* Garden City, N.Y.: Doubleday, 1973.

_____ . *Synergetics*, Vols. 1 and 2. New York: Macmillan, 1975 and 1982.

_____ . *Utopia or Oblivion: Prospects for Humanity* New York: Bantam Books, 1969.

Gadamer, Hans-Georg. *Truth and Method* [1960], trans. Garrett Barden and John Cumming. New York: Seabury Press, 1975.

Galilei, Galileo. *Dialogues Concerning the Two New Sciences* [1638], trans. Henry Crew and Alfonso de Salvio. New York: Dover, 1954.

Gellner, Ernest. *Legitimation of Belief*. Cambridge: Cambridge University Press, 1974.

Gendron, Bernard. *Technology and The Human Condition*. New York: St. Martin's, 1977.

Goldsmith, M. M. *Hobbes's Science of Politics*. New York: Columbia University Press, 1966.

Goodman, Nelson. *Fact, Fiction, and Forecast*, 4th ed. Cambridge: Harvard University Press, 1983.

_____ . *Languages of Art: An Approach to a Theory of Symbols*. Indianapolis: Hackett, 1976.

_____ . *Ways of Worldmaking*. Indianapolis: Hackett, 1978.

Habermas, Jürgen. "Modernity—An Incomplete Project." In *The Anti-Aesthetic: Essays on Postmodern Culture*, ed. Hal Foster. Port Townsend, Wash.: Bay Press, 1983.

_____ . "Philosophy as Stand-In and Interpreter." *After Philosophy: End or Transformation?*, ed. Kenneth Baynes, James Bohman, and Thomas McCarthy. Cambridge: MIT Press, 1987.

_____ . *The Theory of Communicative Action, Vol. I: Reason and the Rationalization of Society*, trans. Thomas McCarthy. Boston: Beacon Press, 1981.

Hacking, Ian. *Representing and Intervening: Introductory Topics in The Philosophy of Natural Science*. Cambridge: Cambridge University Press, 1983.

Harding, Sandra, ed. *Feminism and Methodology*. Bloomington: Indiana University Press, 1987.

Hegel, G. W. F. *The Phenomenology of Mind* [1807], trans. J. B. Baillie. New York: Harper & Row, 1967.

Heidegger, Martin. *Holzwege*. Frankfurt: Klostermann, 1950.

_____ . "The Question Concerning Technology," [1954] In *The Question Concerning Technology and Other Essays*, trans. William Lovitt. New York: Harper & Row, 1977.

Hickman, Larry. *John Dewey's Pragmatic Technology*. Bloomington: Indiana University Press, 1990.

_____ , ed. *Philosophy, Technology, and Human Affairs*. College Station, Tex: IBIS Press, 1985.

Hiskes, Anne L. and Richard P. Hiskes. *Science, Technology, and Policy Decisions*. Boulder, Colo.: Westview Press, 1986.

Hobbes, Thomas. *Leviathan, or the Matter, Forme, & Power of a Common-Wealth Ecclesiasticall and Civill* [1651], ed. C. B. McPherson. New York: Penguin Books, 1968.

Hume, David. *Dialogues Concerning Natural Religion* [1779]. Indianapolis: Bobbs-Merrill, 1947.

_____ . *The Natural History of Religion*, ed. H. E. Root. Stanford, Calif.: Stanford University Press, 1956.

_____ . "Of the Original Contract." In *Political Essays*, ed. Charles W. Hendel. New York: Liberal Arts Press, 1953.

_____ . *A Treatise of Human Nature: Being an Attempt to Introduce the Experimental Method of Reasoning into Moral Subjects*, ed. L. A. Selby-Bigge, 2d ed. Oxford: Oxford University Press, 1978.

Huxley, Aldous. *Brave New World*. New York: Harper & Row, 1969.

Iannone, A. Pablo, ed. *Contemporary Moral Controversies in Technology*. New York: Oxford University Press, 1987.

Ihde, Don. *Existential Technics*. Albany: State University of New York Press, 1983.

_____ . *Technics and Praxis*. Dordrecht: Reidel, 1979.

Illich, Ivan, and Barry Sanders. *A B C: The Alphabetization of the Popular Mind*. San Francisco: North Point Press, 1988.

James, William. *The Meaning of Truth*. Cambridge: Harvard University Press, 1978.

_____ . *Pragmatism*. Cambridge: Harvard University Press, 1978.

Jonas, Hans. *The Imperative of Responsibility: In Search of An Ethics for the Technological Age*. Chicago: University of Chicago Press, 1984.

_____ . *Philosophical Essays: From Ancient Creed to Technological Man*. Englewood Cliffs: Prentice Hall, 1974.

Kant, Immanuel. "The Contest of Faculties." In *Kant's Political Writings*, ed. Hans Reiss. Cambridge: Cambridge University Press, 1970.

_____ . *Critique of Practical Reason and Other Writings in Moral Philosophy*, trans. and ed. Lewis White Beck. Chicago: University of Chicago Press, 1949.

_____ . *Critique of Pure Reason* [1781 1st ed., 1787 2d ed.], trans. Norman Kemp Smith. New York: St. Martin's, 1929.

_____ . *Groundwork of the Metaphysic of Morals*, trans. H. J. Paton. New York: Harper & Row, 1964.

Keller, Evelyn Fox. *Reflections on Gender and Science*. New Haven: Yale University Press, 1985.

Kuhn, Thomas S. *The Structure of Scientific Revolutions*, 2d ed. Chicago: University of Chicago Press, 1970.

Lakatos, Imre, and Alan Musgrave, eds. *Criticism and the Growth of Knowledge*. Cambridge: Cambridge University Press, 1970.

Latour, Bruno, and Steve Woolgar. *Laboratory Life: The Construction of Social Facts*. Princeton: Princeton University Press, 1986.

Lyotard, Jean-François. *Le Différend*. Paris: Éditions de Minuit, 1983.

_____ . *Peregrinations: Law, Form, Event*. New York: Columbia University Press, 1988.

_____ . *The Postmodern Condition: A Report on Knowledge*, trans. Geoff Bennington and Brian Massumi. Minneapolis: University of Minnesota Press, 1984.

_____ . "Presentations." In *Philosophy in France Today*, ed. Alan Montifiore, trans. Kathleen McLaughlin. Cambridge: Cambridge University Press, 1983.

Manuel, Frank E., and Fritzie P. Manuel. *Utopian Thought in the Western World*. Cambridge: Harvard University Press, 1979.

Marcuse, Herbert. "Industrialization and Capitalism in the Work of Max Weber." In *Negations: Essays in Critical Theory*. Boston: Beacon Press, 1968.

_____ . *One-Dimensional Man: Studies in the Ideology of Advanced Industrial Society*. Boston: Beacon Press, 1964.

Marx, Karl. *Capital: A Critique of Political Economy* [1887]. Moscow: Progress Publishers, 1978.

_____ . *Grundrisse: Foundations of the Critique of Political Economy* [1857-58], trans. Martin Nicolaus. New York: Vintage Books, 1973.

Marx, Karl, and Frederick Engels. *Collected Works*. New York: International, 1975.

McMahon, A. P., ed. *Treatise on Painting by Leonardo da Vinci*. Princeton: Princeton University Press, 1956.

Mitcham, Carl, and Robert Mackey, eds. *Philosophy and Technology: Readings in the Philosophical Problems of Technology* 2d ed. New York: Free Press, 1983.

Mitchell, W. J. T. "Pluralism as Dogmatism," *Critical Inquiry* 12 (Spring 1986): 494-502.

Mumford, Lewis. *Technics and Civilization* [1946], 2d ed. New York: Harcourt, Brace and World, 1963.

Neurath, Otto. *Empiricism and Sociology*, eds. Marie Neurath and Robert S. Cohen. Dordrecht: Reidel, 1973.

_____. "Protocol Sentences," *Logical Positivism*, ed. A. J. Ayer. Boston: Free Press, 1959.

Nietzsche, Friedrich. *The Birth of Tragedy Out of the Spirit of Music* [1872], trans. Walter Kaufmann. New York: Vintage Books, 1967.

_____. *Ecce Homo* [1888], trans. Walter Kaufmann. New York: Vintage Books, 1969.

_____. *The Gay Science* [1882], trans. Walter Kaufmann. New York: Vintage Books, 1974.

_____. *On The Genealogy of Morals* [1887], trans. Walter Kaufmann. New York: Vintage Books, 1967.

_____. "On Truth and Falsity in an Extra-Moral Sense," [1873] In *Complete Works of Friedrich Nietzsche*, Vol. 1, trans. Maximilian Mugge, ed. Oscar Levi. New York: Russell and Russell, 1964.

_____. *Thus Spoke Zarathustra* [1883-84], trans. Walter Kaufmann. New York: The Viking Press, 1966.

_____. *The Use and Abuse of History* [1874], trans. Adrian Collins. Indianapolis: Bobbs-Merrill, 1957.

_____. *The Will to Power*, trans. Walter Kaufmann and R. J. Hollingdale, ed. Walter Kaufmann. New York: Vintage Books, 1967.

Ormiston, Gayle L., ed. *From Artifact to Habitat: Studies in the Critical Engagement of Technology*. Bethlehem, Penn.: Lehigh University Press, 1990.

Ormiston, Gayle L., and Alan D. Schrift, eds. *The Hermeneutic Tradition: From Ast to Ricoeur*. Albany: State University of New York Press, 1989.

_____, eds. *Transforming the Hermeneutic Context: From Nietzsche to Nancy*. Albany: State University of New York Press, 1989.

Orwell, George. *1984*. New York: New American Library, 1981.

Pacey, Arnold. *The Culture of Technology*. Cambridge: MIT Press, 1986.

_____. *The Maze of Ingenuity: Ideas and Idealism in the Development of Technology* [1974] Cambridge: MIT Press, 1985.

Peirce, Charles Sanders. *Collected Papers*, ed. Paul Weiss and Charles Hartshorne. Cambridge: Harvard University Press, 1931-35.

Popper, Karl R. *Conjectures and Refutations: The Growth of Scientific Knowledge*. New York: Harper & Row, 1963.

_____. *The Logic of Scientific Discovery* [1934]. New York: Harper & Row, 1968.

Rapp, Friedrich. "Philosophy of Technology." In *Contemporary Philosophy: A New Survey*, vol. 2, ed. Güttorum Floisfod. The Hague: Martinus Nijhoff, 1982.

Richter, Irma A., ed. *Notebooks of Leonard da Vinci*. Oxford: Oxford University Press, 1982.

Richter, Maurice N., Jr. *Technology and Social Complexity*. Albany: State University of New York Press, 1982.

Rorty, Richard. *Consequences of Pragmatism*. Minneapolis: University of Minnesota Press, 1982.

_____ . *Philosophy and the Mirror of Nature*. Princeton: Princeton University Press, 1979.

Rothschild, Joan. *Teaching Technology From a Feminist Perspective*. Oxford: Pergamon, 1988.

Rothschild, Joan, ed. *Machina ex Dea: Feminist Perspectives on Technology*. New York: Pergamon, 1983.

Rousseau, Jean-Jacques. *Emile or On Education*, trans. Allan Bloom. New York: Basic Books, 1979.

_____ . *The First and Second Discourses*, ed. Roger D. Masters. New York: St. Martin's, 1964.

Ryle, Gilbert. *The Concept of Mind*. New York: Barnes and Noble, 1949.

Sassower, Raphael. *Philosophy of Economics: A Critique of Demarcation*. Lanham, Md.: University Press of America, 1985.

Schrader-Frechette, K. S. *Nuclear Power and Public Policy: The Social and Ethical Problems of Fission Technology*. Dordrecht: Reidel, 1980.

_____ . "The Real Risks of Risk-Cost-Benefit Analysis." *Technology and Responsibility*, ed. Paul T. Durbin. Dordrecht: Reidel, 1987.

Schumacher, E. F. *Small is Beautiful: Economics as if People Mattered*. New York: Harper & Row, 1973.

Silby, Mulford Q. *Technology and Utopian Thought*. Minneapolis: Burgess, 1971.

Snow, C.P. *The Two Cultures*. Cambridge and London: Cambridge University Press, 1964.

Spragens, Thomas A. Jr. *The Politics of Motion: The World of Thomas Hobbes*. Lexington, Ken.: University Press of Kentucky, 1973.

Urbach, Peter. *Francis Bacon's Philosophy of Science: An Account and a Reappraisal*. La Salle, Ill.: Open Court, 1987.

Volti, Rudi. *Society and Technological Change*. New York: St. Martin's, 1988.

Winner, Langdon. *Autonomous Technology: Technics-Out-of-Control as a Theme in Political Thought*. Cambridge: MIT Press, 1977.

_____ . *The Whale and the Reactor*. Chicago: University of Chicago Press, 1986.

Wittgenstein, Ludwig. *Philosophical Investigations*, trans. G. E. M. Anscombe. New York: Macmillan, 1968.

Woolgar, Steve. *Science: The Very Idea*. London and New York: Tavistock, 1988.

Yearley, Steven. *Science, Technology and Social Change*. London: Unwin Hyman, 1988.

Xenophon. *Oeconomicus*, trans. C. Lord In Strauss, Leo. *Xenophon's Socratic Discourse*. Ithaca: Cornell University Press, 1970.

Zamiatin, Eugene. *We*, trans. Gregory Zilboorg. New York: Dutton, 1952.

INDEX

INDEX

Gayle L. Ormiston is associate professor in the department of philosophy and the Institute for Applied Linguistics at Kent State University. He has also taught at the University of Colorado at Colorado Springs and at Denison University. Ormiston received his B.A. and M.A. in philosophy from Kent State University and his Ph.D. in philosophy from Purdue University. His articles have appeared in *Semiotica, Ars Semeiotica, Philosophy Today, The Journal of Aesthetics and Art Criticism*, and *Journal of the British Society for Phenomenology*. Ormiston is the editor of *From Artifact to Habitat: Studies in the Critical Engagement of Technology* (1990) and co-editor of *Transforming the Hermeneutic Context: From Nietzsche to Nancy* (1989) and *The Hermeneutic Tradition: From Ast to Ricoeur* 1989).

Raphael Sassower is assistant professor in the department of philosophy at the University of Colorado at Colorado Springs. Sassower received his M.A. and Ph.D. in philosophy from Boston University. He previously taught at the University of Massachusetts. Sassower is the author of *Philosophy of Economics: A Critique of Demarcation* (1985) and contributes to *Philosophy of the Social Sciences, Social Epistemology, Theoretical Medicine*, and the *Journal of Business Ethics*.